TABLE OF CONTENTS

Ordering information
800-927-8800 phone
206-232-8800 phone
206-232-2741 fax

ACKNOWLEDGEMENTS

Special thanks to Diane, Larry and Steve for their encouragement throughout this project. I appreciate Fred's unlimited talent that created the unique cover. My hat goes off to my assistant, Julie, for her continued support and ideas. Scott, my 6-year-old son, made significant contributions in the creativity department.

THE BIG PICTURE

The year is 2010. Driving down the road, my car sputters, then stalls. I immediately pull to the side of the road and call my mechanic from my mobile phone. He accesses the cellular network using his station-to-vehicle communications center and runs diagnostics on my broken-down vehicle.

"I'm afraid you need a new water pump," he replies, "I have one in supply and can bring it to you. I see from my vehicle-tracking system that you are at the corner of Main and First Streets. We will have you on the road again in no time."

I mutter to myself, "They can build a mobile communications network that can take my phone messages, activate my lights at home and provide an alarm system for my home and car, but they can't build a car that won't break down when I need it most."

Although this vision is purely conjectural, it is not only highly possible, but quite likely. The cellular network of today is, and will be, but a small part of the mobile communications picture of tomorrow. From one good idea an industry springs, and from that industry, new peripheral technologies arise. It has happened with other technologies and is bound to happen with cellular.

In 1863, when Henry Ford presented his first automobile design, he might have laughed if someone had told him that his "horseless carriage" would turn into a multibillion dollar industry, with automobile manufacturers around the world, gas stations on every street corner and major highway systems to support those vehicles. And if Watson had told Alexander Graham Bell that within the century there would be hundreds of millions of telephones around the world, he might have hung up on him. And if someone had told our parent's parents that their grandchildren would be running offices out of their cars, they might have wondered if someone had been dipping into the moonshine.

The exciting part of the cellular industry is that the potential of cellular and wireless communications is limitless. Take a minute to look beyond the next decade. You can almost see a world in which people no longer use phones plugged into the wall. Wireless handsets are as common as chocolate on Valentine's Day. People will receive and transmit data to and from sources as easily as we change a radio station in our cars today. And all this thanks to what started quietly in 1983.

What began as a fledgling industry in Chicago has boomed into a multibillion dollar industry. And it is growing. We are still moving at breakneck speed. However, instead of moving in one direction of only selling mobiles, transportables and portables to budding executives, cellular is moving in all directions. Cellular is used in fixed locations in homes and offices around the world, with credit cards and in emergency medical services. It is used for backup service to landlines and as non-detectable burglar alarms. And the future promises more interesting applications, such as remote vehicle diagnostics and enhanced vehicle-tracking systems.

Looking ahead is an entertaining mental exercise, but we must remember where we came from as well. That is where all of this dreaming started. And by understanding the basic principles, we will be able to keep pace with this industry that is progressing so rapidly.

It is easy to get left behind in an industry that moves so quickly. Fortunately, today there are publications and books about the cellular industry that provide the big picture, that show you what is beyond your company. Don't let your knowledge stop at your office door. Today you have at your disposal more information than ever to enhance your cellular knowledge and understanding.

We are all part of this leading edge of telecommunications technology for the next century. We are showing the world that it is possible to be reached wherever you are, not wherever your phone is plugged into a wall. This is your opportunity to become an important part of the cellular communications picture. You can then share that knowledge with your peers and with your customers. And by doing so, you will become a complete cellular professional who **is** part of the big picture.

Rhonda L. Wickham
Editor, *Cellular Business*

INTRODUCTION

The greatest challenge of this book has been to create a document that everyone can understand. Throughout the years, I have conducted several workshops in the cellular industry. The participants have included technicians, customer service representatives, installers and salespeople. This book attempts to deliver a concise and informative product that makes everyone's job a little easier.

My cellular work history includes being a service manager, installer, retail manager and trainer. Over the years, several cellular "buzz" words and ideas became second nature. It is easy to use words like NAM or directed retry with cellular freshmen and customers. Unfortunately, this type of talk with newcomers can be very frustrating for both parties.

The general purpose of this book is to explain many aspects of cellular in simple terms. I have learned that subjects we are closest to are most difficult to explain. During the writing of *Simplified Cellular*, I used a software program that checked my writing for grammar and vocabulary. It is similar to having an English teacher on staff without the frivolous arguments. I was amazed how a simple chapter, to me, was rated as "too technical". Fortunately, after several arguments with the computer, the book was written.

I have also incorporated plenty of graphics to help deliver the messages throughout the book. Some graphics may seem a bit silly, but please remember, my need for personal entertainment is at the same level as food and sleep.

Part 1

How It Works

Cellular Evolution

The Network

Roaming

1 Cellular Evolution

The Initial Cellular Test

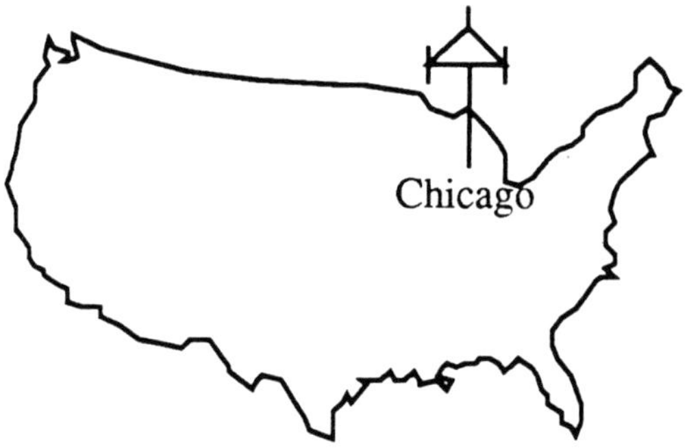

Cellular service began with a test system in Chicago. The cellular engineers introduced a wonderful idea in mobile communications. The previous phone systems used a single base antenna with limited channels. It was common for a caller to be denied access to pre-cellular telephone systems because of system overload. Network growth limitations required subscribers to wait several months to be assigned a telephone number.

Unlike its predecessor, the new cellular design opened a large untapped market for mobile communications. Cellular systems operate with several cells in the network and reuse the channels in different cells. Because of cellular's low operating power, channels can be reused in non-adjacent cells. For example, a channel in a downtown cell also may be used in an industrial area, several miles away.

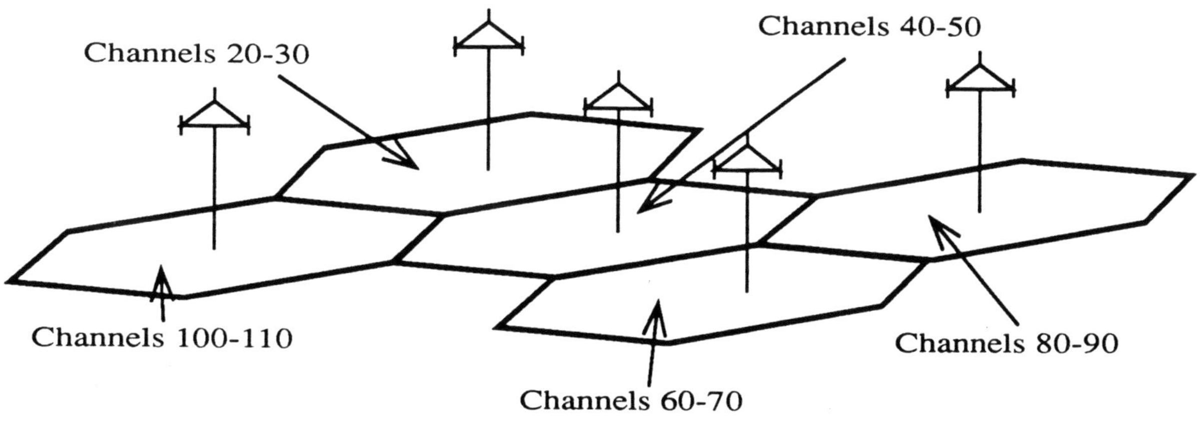

The channels and their orchestration are handled by the system. Cellular customers have no idea which channel the phone may be tuned to. The system assigns the mobile or portable a specific cell and channel within the system. A call originates in one cell and passes from cell to cell as the user travels in the cellular system. To put it simply, it's a nifty idea.

The original cellular mobiles were bulky and adhered to the Advanced Mobile Phone Service (AMPS) standards. These standards specified the phone's power output, method of programming and cabling design. AMPS criteria also specified the phone's operating frequencies and method of communicating with the network.

The United States and many other countries continue to use the AMPS standard for cellular systems. Phone manufacturers continued to use many original AMPS specifications. Most phone manufacturers have varied from some AMPS requirements, such as data cable design and NAM storage methods (used to program phones).

In the Chicago test, mobiles were huge in comparison to phones manufactured today. Over the past several years, the phones have shrunk in size and now offer additional features. The postage-stamp phone with 3,000 features could be the phone of the future.

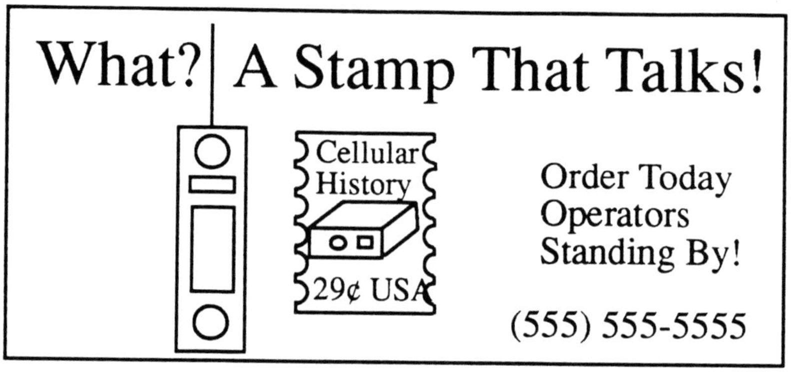

Once the Chicago test was complete, the Federal Communications Commission (FCC) accepted the cellular concept. In 1982, systems began to appear throughout the United States.

Two Systems per Market

To promote a competitive marketplace, the FCC granted two licenses in each cellular market. Cellular first sprouted in the 30 largest U.S. cities. Licenses were granted to the smaller markets in order of population until all major markets were part of a nationwide cellular network. The areas between major markets on interstate highways and small towns, called rural service areas (RSA), were the last to get cellular service. In summary, MSAs first, RSAs next and the bare spots with no population received no cellular service.

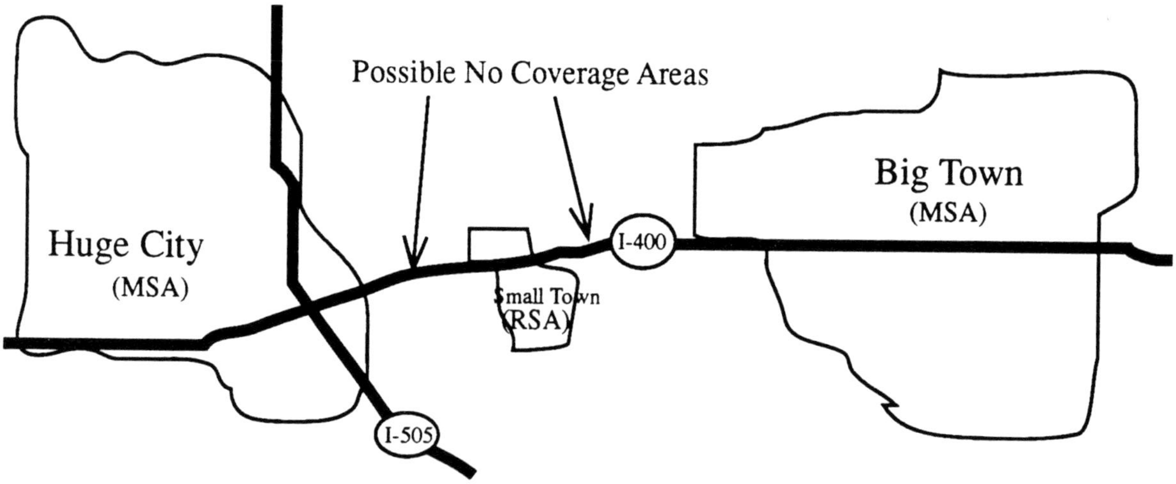

The pecking order of larger markets first and smaller markets next was no accident. Building a cellular system is costly and time consuming. The early 1980's cellular technology required a substantial customer base to meet the expenses. Cellular systems had large up-front costs to be absorbed by the cellular customers.

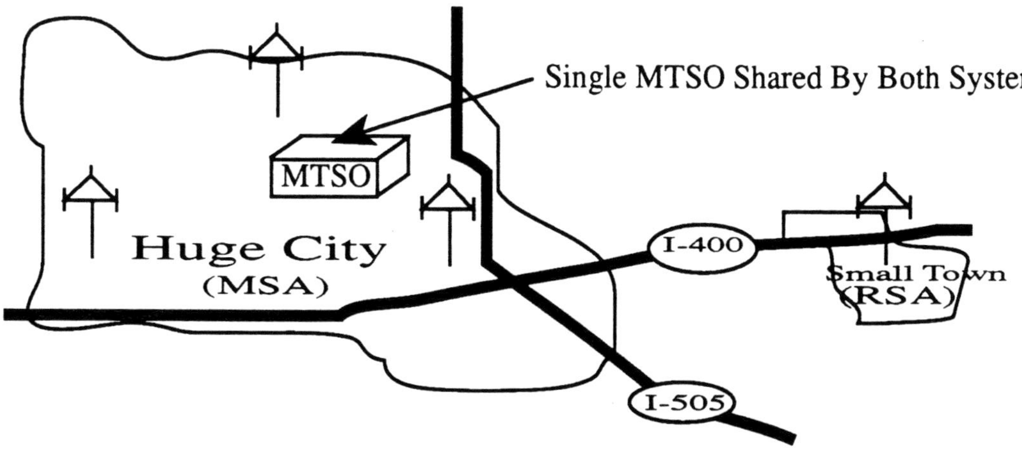

As the 80's progressed, the cost of installing a cellular antenna system and switching office decreased. This helped make cellular possible for markets with fewer customers. To lessen the financial burden, some small markets shared their mobile telephone switching office (MTSO) with nearby existing cellular systems. A system "back hauls" if it uses another network's MTSO. A small MSA or RSA may back haul several hundred miles to save the cost of a switch. The carriers determined that the long distance charges were cheaper than the price of a second switch.

Cellular Channels

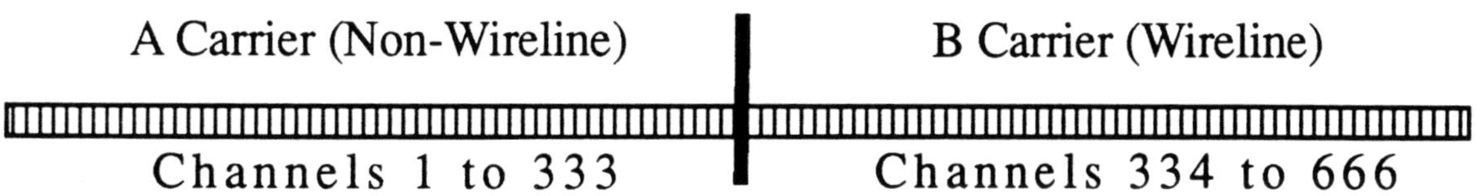

666 Channel Cellular System

The original cellular systems operated with 666 channels. These channels provided carrier A (non-wireline) with 333 channels and carrier B (wireline) with the remaining 333 channels. Carrier A used channels 1 through 333 and carrier B operated with channels 334 through 666.

The wireline operator originally was a distant cousin of the local phone company and the non-wireline carrier was funded by private investors. One myth during cellular's early days was that the wireline systems connected the system's cells to the switching office with wires or fiber optic lines. Many people believed that the non-wireline carrier used microwave connections between the cells. It must have sounded believable because the rumor continued for years. In reality, both systems use a combination of telephone lines and microwave interconnects in building a system.

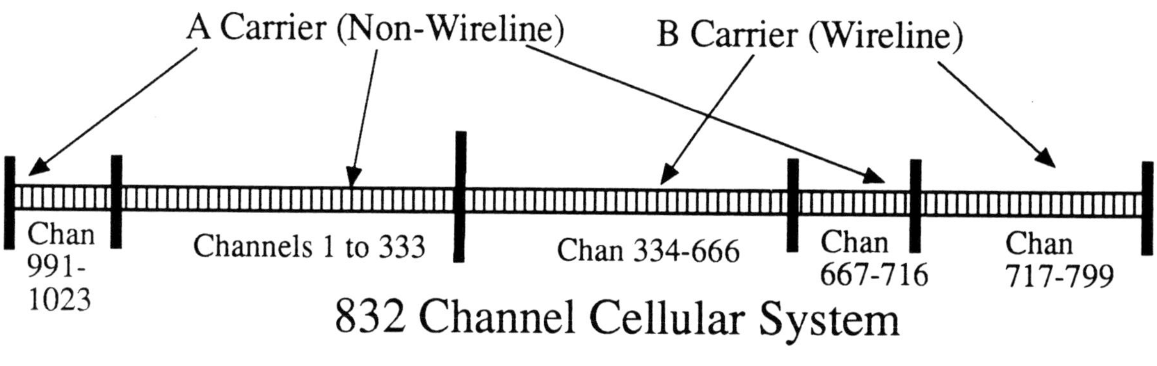

As the number of subscribers grew, several densely populated systems outgrew the 666 channel maximum. Cellular networks obtained expanded spectrums in 1987, increasing the total cellular channels to 832. This new channel allocation increased each carrier's usable channels from the previous 333 to the current 416.

What Happens to a 666 Channel Phone in an 832 Channel System?

The giant leap to additional channels took some awkward steps. The early phones were designed to operate on a 666 channel system. During the activation phase, technicians program specific system parameters into the phone. One parameter, station class, specifies if the phone can operate on the expanded spectrum. Some 666 channel phones were mis-programmed during the transition to the added spectrum. The mis-programming told the network that the phone could handle the upper channels (above 666).

A mis-programmed 666 channel phone can have multiple dropped calls

The subscriber would return to the shop with complaints that the phone would not complete calls or that conversations were cut off during a hand-off to another cell. Fortunately the carriers, in a proactive effort, identified most of the problem phones by watching for excessive dropped calls, (a disconnect not initiated by the mobile). Carriers eliminated the problem once they identified the station class culprit.

Phones manufactured before the expanded spectrum operate satisfactory in an 832 channel system if programmed correctly. Cellular networks never assign a properly programmed 666 channel to expanded spectrum channels. From the user's perspective, both 666 and 832 channel phones seem to act the same. There is no difference in the phone's audio quality or call processing ability.

Global Cellular

Several European countries introduced cellular before the cellular systems in the United States. These systems had unique channel spacing and operating frequencies. This made many European systems incompatible with the AMPS system in the United States.

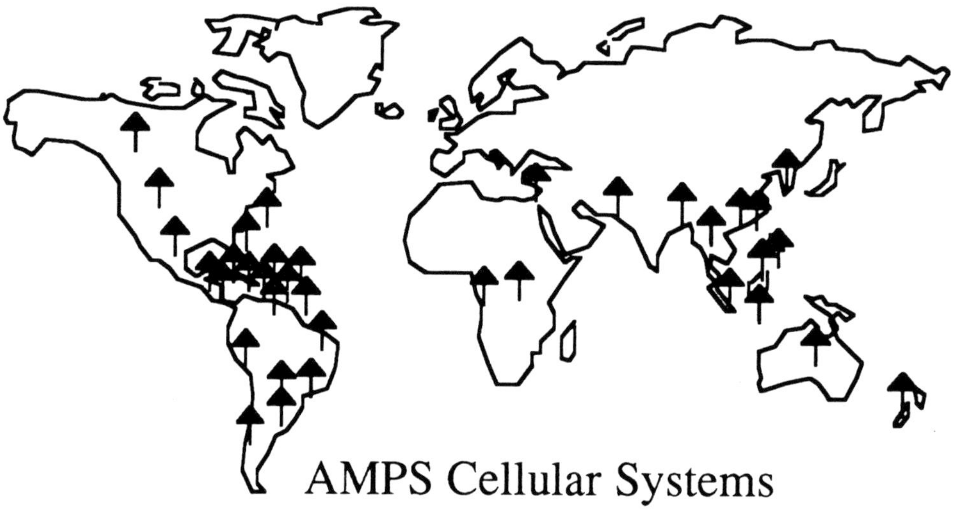

AMPS Cellular Systems

As the United States built its AMPS cellular system, several other countries had the same dream. Fortunately our neighbors, Canada and Mexico, built AMPS compatible systems. This simplified customer roaming in the neighboring systems and provided instant customers for the switch and phone manufacturers.

Countries operating with systems not compatible with AMPS specifications include England, Sweden and Japan. Some systems currently using the AMPS standard include Mexico, Canada, Puerto Rico, Australia, New Zealand, Virgin Islands, Bahamas and the Philippines. The list of AMPS compatible systems is growing at a healthy rate. This is great news for U.S. cellular users traveling outside the country.

AMPS Compatible Cellular Systems
(Courtesy of Communications Publishing, Mercer Island Washington)

Antigua	Mexico
Argentina	Netherlands Antilles
Australia	New Zealand
Bahamas	Pakistan
Bangladesh	Paraguay
Bermuda	Peru
Brazil	Philippines
Brunei	Puerto Rico
Canada	St. Kitts and Nevis
Cayman Islands	St. Lucia
Chili	Samoa
Costa Rica	Singapore
Dominican Republic	South Korea
Gabon	Taiwan
Grenada	Thailand
Guatemala	United States
Hong Kong	Uruguay
Indonesia	Venezuela
Israel	Virgin Islands
Jamaica	Zaire

2 The Network

SYSTEM OPERATION

Cellular systems have four basic components. These include the mobile telephone switching office (MTSO), individual cells, mobile phones (car phones and portables) and an interconnect into the local telephone company. The drawing below illustrates the cellular system elements.

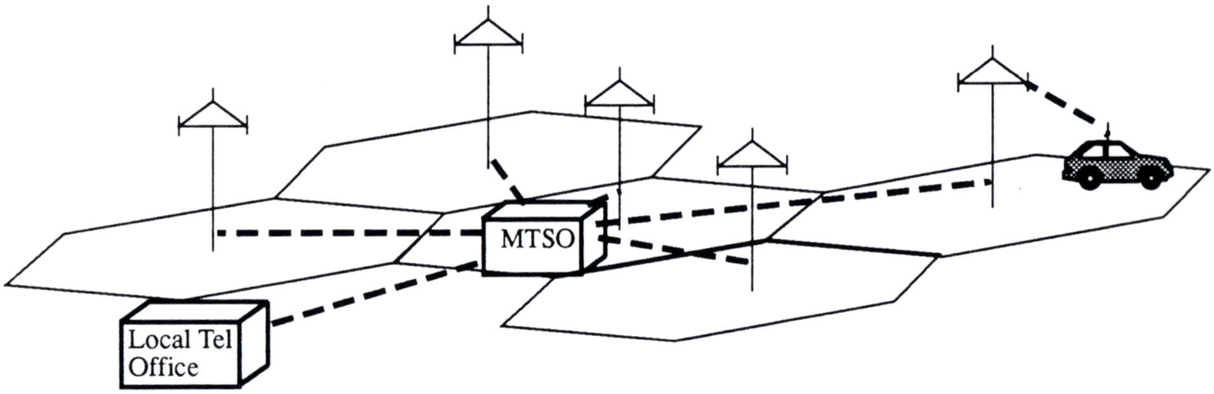

Cellular is often graphically shown as a series of honeycomb or hexagon cells. In reality, the neatly organized hexagons are a series of free-form circular shapes that either overlap or miss each other completely, causing "holes" in the cellular coverage.

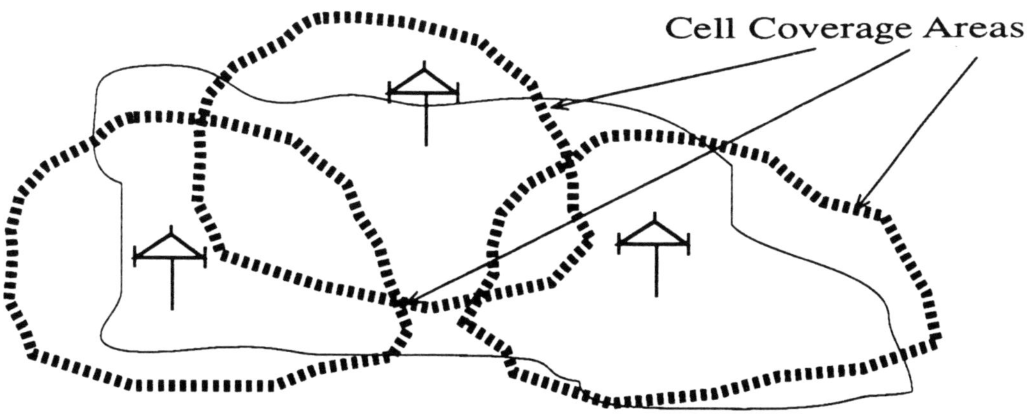

The cellular idea involves low power mobiles for the users and low power transmitters at the cell sites. Since both the mobiles and cells operate at a low power, the coverage of each is approximately 12 miles. This figure can vary considerably depending on terrain such as water, which extends coverage or hills, which limit range. Another coverage factor is folliage. Areas with plenty of broad-leafed trees may suffer from reduced system coverage.

Cellular systems usually have multiple cells, and as the network matures, channel reuse is common. Because of the system's low power design, channel reuse is possible with a two-cell separation. In other words, channel number 500 can be used in multiple cells. To prevent multiple calls from being made on channel 500, the cells using the same channel must be separated by at least two cells. A two-cell separation often exceeds 30 miles.

Cells

While operating within a cell, the cellular system assigns the mobile a channel in the closest cell. As the mobile travels from one cell to the next, the system assigns another channel in an adjoining cell. The system commands the phone to "handoff" to the adjacent cell as necessary.

System usage will dictate the number of channels assigned to a specific cell. A busy cell may contain more than 50 channels, while a cell in a remote area may have only 5 channels. The cellular system engineers monitor a cell's usage to decide if additional channels should be installed in an existing cell. A lively part of the system may justify adding an additional cell. System activity and customer feedback have a tremendous effect on system improvements.

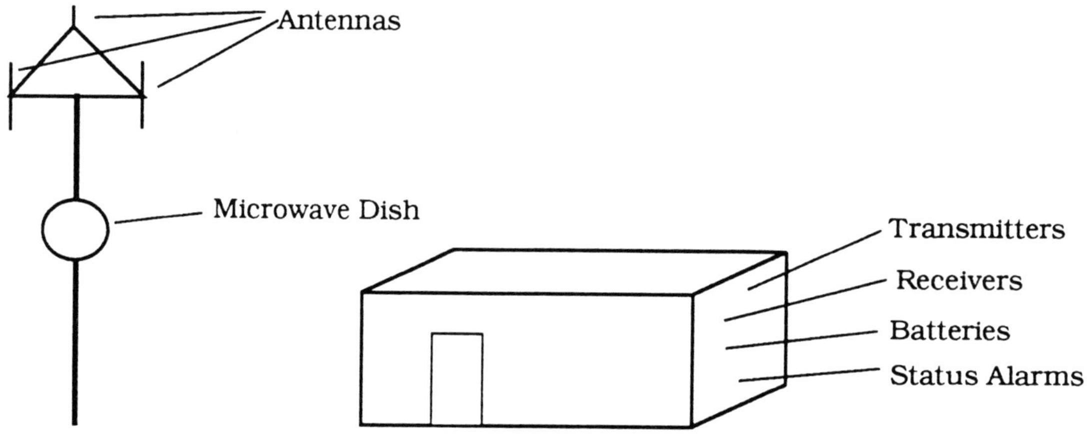

Cell Site Components

The cells contain transmitters and receivers that talk with the cellular telephones operating within the system. Cells also have status alarms and component redundancy for maximum reliability. Alarms monitor the cell's temperature, humidity, fire, defective channels and burglary. Cell alarms alert the MTSO technicians of potential problems at the cell.

Mobile Telephone Switching Office (MTSO)

The most powerful system component is the MTSO. The MTSO houses the computer controls for the system.

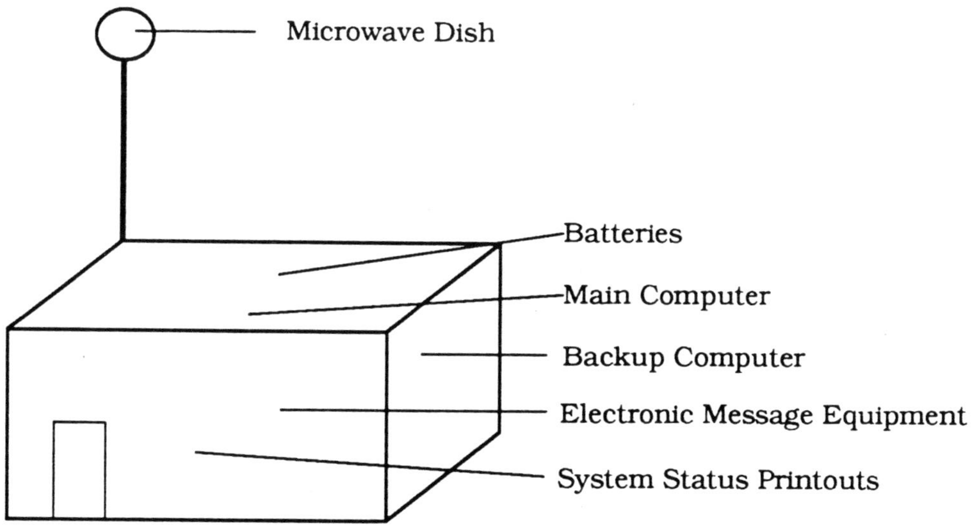

MTSO Components

The system has multiple computers running in parallel. Redundancy is evident with multiple computers, alarms and channels. If one computer should "crash", the second can seize the system controls. The MTSO is responsible for many tasks in the system. MTSO functions include call set up, channel assignment, mobile power level adjustment, cell alarm monitoring and call termination. The cells and user equipment can be viewed as machines operating per MTSO demands. The MTSO is like a friendly dictator.

Subscriber Equipment (Mobile and Portable Phones)

Operating within this high-tech maze is the customer's equipment. Mobile and portable phones can place and receive telephone calls as permitted by the system. The cellular customer can place a call request, yet the fate of call completion is dependent on several factors. Once the mobile makes the call request, the MTSO assigns the mobile an available channel at a neighboring cell site. If a handoff is necessary, the MTSO arranges and orchestrates the cell change for the mobile. The point is that the mobile operates per instructions from the cellular system.

Telephone Company Interconnect

The final piece of the cellular system puzzle is the telephone company interconnect. The outgoing signal from a customer's cellular phone can be traced through the system. The signal transmits from the vehicle to a nearby cell. From the cell, the signal travels to the MTSO via telephone lines or a microwave system. The processed signal arrives at the MTSO where it must be connected to the landline telephone system. The connection between the MTSO and the local telephone office is a series of telephone lines. Once the mobile's signal reaches the local telephone office, the call proceeds like a call originating from your home or office.

Cell Channels

The cells have two types of channels; one sends data (like the sound your fax machine makes), and the other processes data and the audio. The first channel type is the data channel. The second is a voice and data combination channel. Each type of channel has a specific function in the system.

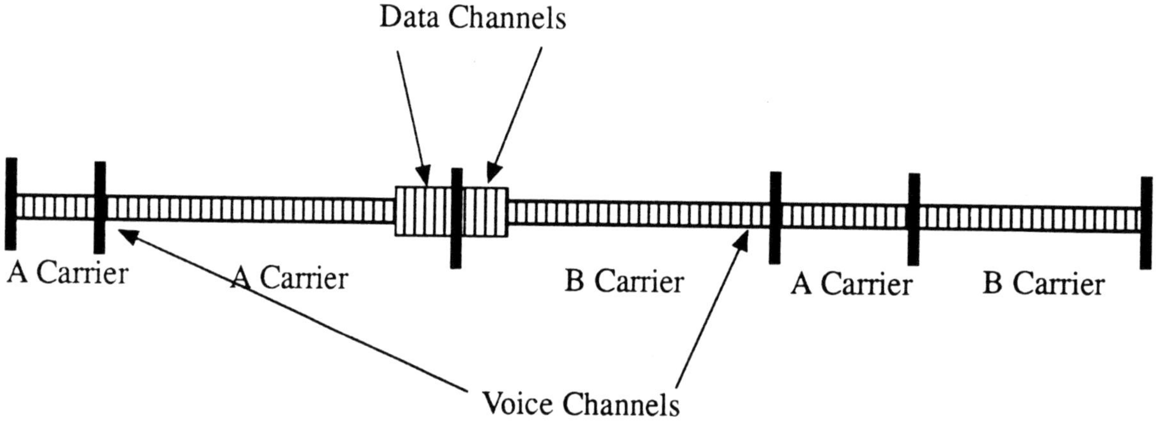

The system and mobile use the data channels to transmit information to each other with no call in progress. Data includes system parameters, such as the system information, incoming pages (call notifications) to mobiles and occasional requests for a phone to identify its name, rank and serial number.

Data also travels from the mobile to the cell on the data channel. This data tells the cellular system to originate a call and assign a voice channel. Once the system or the mobile requests a call initiation, the system assigns the mobile to a second type of channel, the voice channel. Voice channels transmit both data (digital) and audio (analog) signals. This may seem confusing. Why mix the whir of a fax machine and a conversation with your stock broker? The phone filters the data; thus the cellular user never hears the data interruptions. The system uses data to tell the phones to switch cells, or to adjust power levels.

Autonomous Registration

Many cellular systems have a feature known as autonomous registration. Autonomous registration tells the system the location of phones operating in the system.

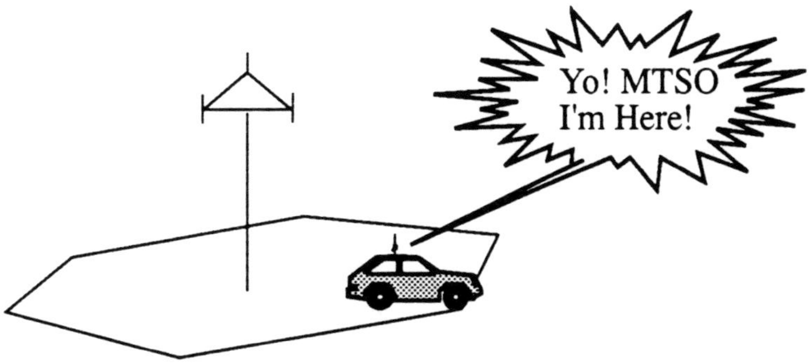

Autonomous Registration

Without autonomous registration, a cellular phone receiving an incoming call must be paged throughout the cellular system. By having a location map of current mobiles, the system shortens the paging time. With autonomous registration, the network only pages registered mobiles within specific geographic areas. This saves work for the system and expedites call processing for the cellular user.

Electronic Serial Number

Each cellular telephone has a discrete electronic serial number (ESN). The phone manufacturers program the ESN at the phone's birth and an ESN cannot be legally altered by anyone other than the phone manufacturer.

When a user places a call, the system's computer verifies the programmed mobile number and the ESN transmitted to the cellular system. If the transmitted numbers don't match the system's data base, the network denies the call. Only one ESN can be assigned to a specific cellular telephone number, thus making it impossible for multiple phones to share one mobile number. The solution for customers desiring multiple phones with a single telephone number is call forwarding. With call forwarding a customer can distribute the phone number for one phone. Calls to the published number can be forwarded to any other landline or cellular telephone.

Placing a Call

Cellular telephones initialize when powered "on". The term "initialize" refers to the phone performing a series of self-checks before allowing calls to be placed. This step is similar to the pre-check an airplane pilot goes through before flying the plane. The phone checks various internal modules and the programming contents.

If everything is within specifications, most phones display a series of 8s and a short tone sounds. The phone's No Service light illuminates temporarily while the telephone's receiver section listens for the strongest data channel, also called a control channel. The phone's goal is to find the closest cell to establish a data link.

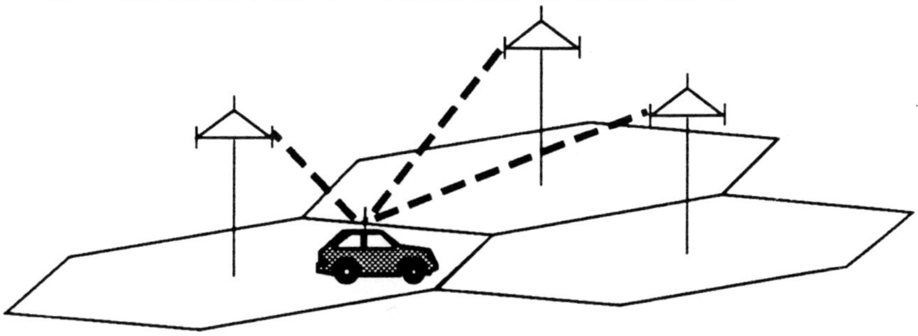

Phone scans for strongest data channel

Since each cell has only one control channel to page mobiles, the mobile can scan for the strongest control channel. The strongest channel generally means the closest cell. This channel transmits network data to the mobile such as the system identification number, control channels used and incoming call alert. The data channel also can receive information from the mobile, such as a request for the mobile to place a call.

Once the initialization process is complete and the phone's No Service light extinguishes, calls can be placed. To place a call, the cellular user dials a phone number from the keypad or recalls a telephone number from a memory location. The user presses the SEND button and initiates a call request to the network.

Cellular phones never emit a dial tone from the handset. Before pressing the send key, the phone is in the idle mode listening to a data channel. The phone requests a call set up to the system on the same control channel. The cellular system records the number dialed, the mobile's phone number and the phone's ESN. If the system recognizes the ESN and mobile number, the call process continues.

The MTSO assigns the mobile a voice channel in the same cell as the data channel used for the call origination. Once the phone jumps to the voice channel, the user hears the phone ringer to suggest the call has been connected to the dialed party. The signal from the mobile travels from mobile to cell, cell to MTSO, MTSO to local telephone office and finally arrives at the landline telephone. It is amazing this complicated process happens in seconds.

If the cellular user presses the END button, the call is disconnected. The END key causes the cellular phone to emit a short, high-pitched tone to tell the system to end the call. The user and called party do not hear the extra tone because of the phone's filtering system.

Receiving a Call

Incoming calls have similar steps to outgoing telephone calls. The phone powers "on" and scans, looking for the strongest control channel. The phone locks on the control channel, causing the No Service light to extinguish. It listens for system information and incoming pages. The term "pages" refers to the network's method of notifying the phone of an incoming telephone call. Once the network sends the page to the cellular phone, the phone responds on the same control channel. It notifies the system of the phone's presence and cell location.

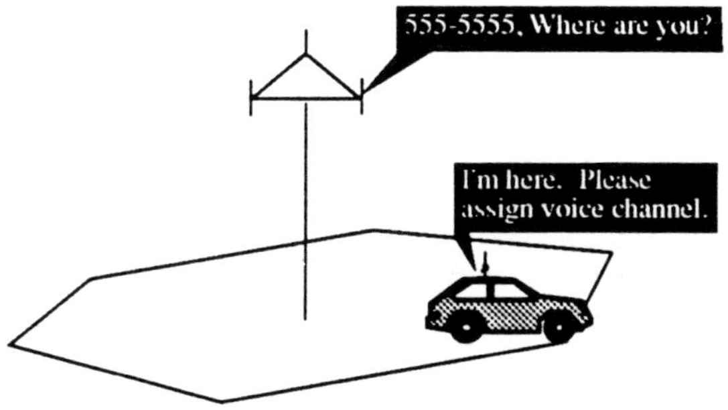

Data Channel Information

By acknowledging the system, the MTSO can assign the phone a voice channel at the closest cell site. The phone leaps to the assigned channel and the cellular phone starts to ring. The concept of control channels and voice channels is a critical element to understanding the call set up process. An easy way to remember the function of each type of channel is voice channels have voice and data/control channels send and receive data.

The Handoff

With a working knowledge of making and receiving calls we can now tackle handoffs. The term "handoff" describes the process of mobiles transferring from one cell to another. Cellular network design involves a series of low-power transmitters in each cell. Since the coverage of each cell is limited, the system must be able to transfer the call. As the phone travels from cell to cell, the system orchestrates the handoff process to continue the call.

Let's assume the customer is on a voice channel with a call in progress. The network monitors the call and notices the noise level is increasing to the point of sounding scratchy to the user. The adjoining cells monitor the mobile's signal to figure out which cell is closest.

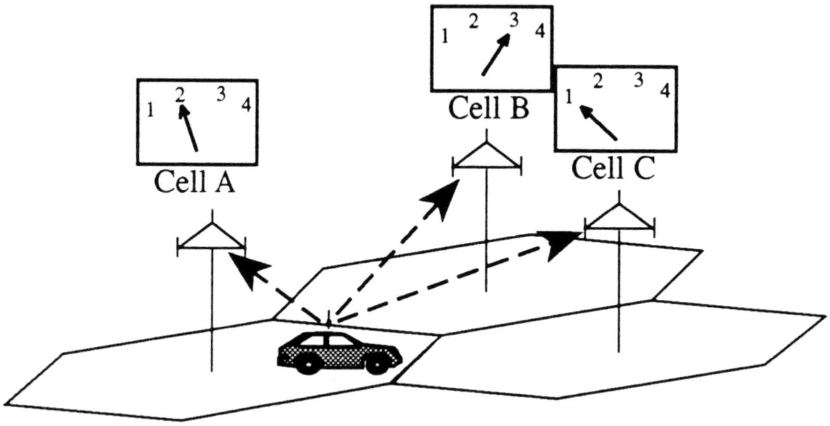

System monitors mobile signal to determine closest cell

The current cell (A) may show a signal level 2, another cell (B) may measure a signal 3 and the last cell (C) may see a signal of 1. From the sampling, the system knows the best cell for the customer is cell (B) with the clearest signal. The network assigns the mobile a voice channel on cell (B) and the cellular phone jumps to cell (B), channel XYZ. The handoff process is complete in a fraction of a second and generally the user is unaware of the channel change.

Leaving a Cellular System

A mobile traveling out of a cellular system experiences what I call "going, going, gone." The system monitors calls and decides if a handoff is necessary. Since the phone is traveling away from the system, the adjoining cells show a weaker signal than the cell currently in use.

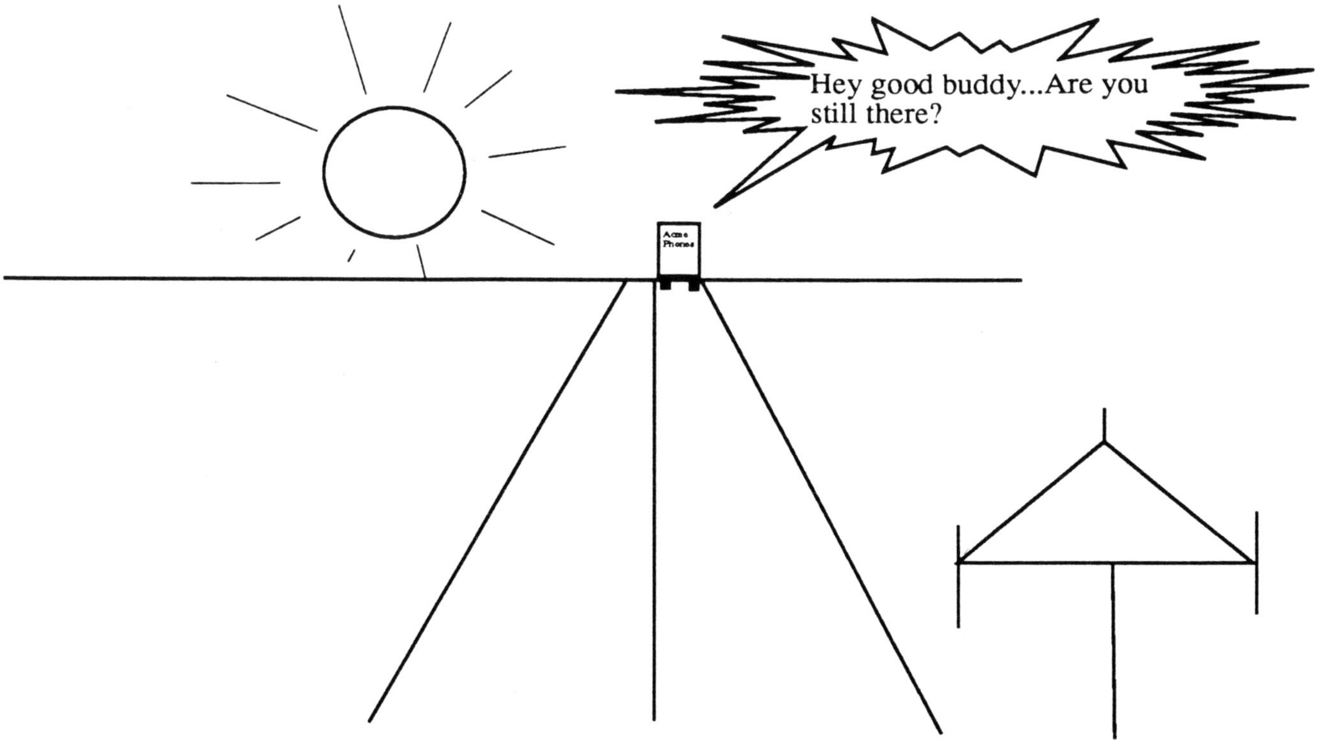

The frustrations of driving out of cellular coverage

The conversation may go quiet for 2 seconds then pop back for a few seconds. The next step is a 3-second break in the conversation. If the network loses the conversation for 5 seconds straight, the call ends. This can be a blessing or a headache, depending on who you're talking to on your phone. The reason for the ended call, after 5 seconds, is cellular carriers charge for calls by the minute. If systems didn't have this safeguard, calls would never end unless the user pressed the END button. Could you imagine the cellular bill if the airtime continued and you left town for a week at $X.XX per minute? The reason for the second mortgage would leave your banker in hysterics.

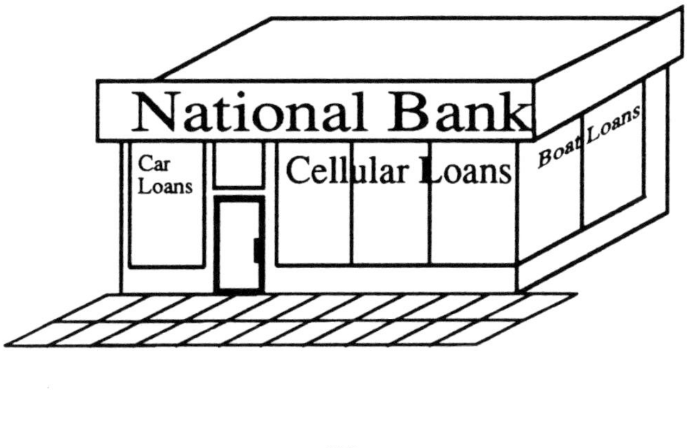

The Cellular Spectrum

The diagram below illustrates the frequencies used by AMPS cellular networks.

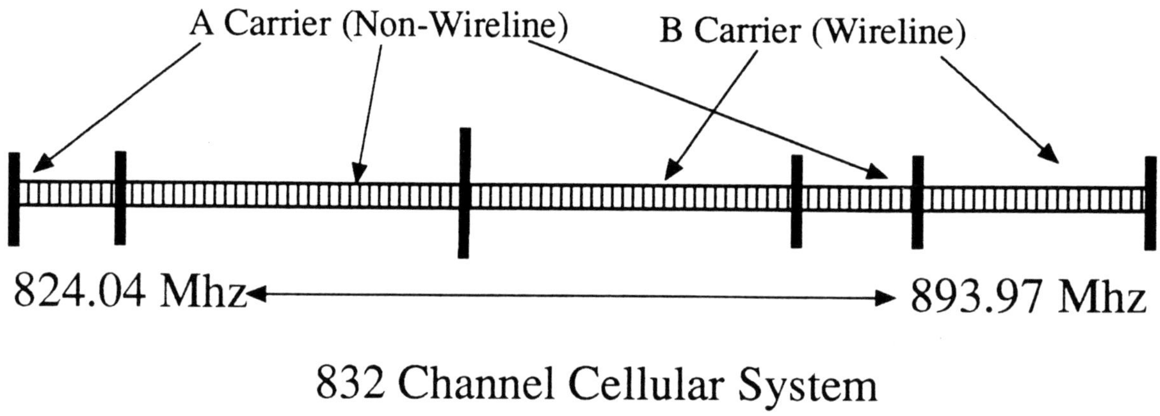

Cellular control and voice channels reside in the 800 megahertz (Mhz) range. The term "megahertz" may sound confusing until you understand that "mega" means million and "hertz" stands for cycles per second. The cellular spectrum is in the 800,000,000 cycles-per-second range. Your house current is 60 cycles per second. In other words, 800 Mhz is a high frequency.

The FCC divided the cellular frequencies into two channel sets. The A carrier (non-wireline) uses one set and the B carrier (wireline) occupies the second set. These channel sets can be divided again into voice and control channels. A phone in the idle mode waits and listens to a data channel. The phone connects to a voice channel only during a call.

Analog and Digital Cellular

Cellular systems in the mid 1980s had 666 channels. The A and B carriers in each market divided the channels with 333 channels for each carrier. In cities such as New York and Los Angeles, 666 channels couldn't handle the call load placed on the system. The

Federal Communications Commission (FCC) allotted more channels in 1987 to rectify the problem. This brought the total to 832 channels, split evenly between the A and B carriers. As the subscribers continued to grow, 832 channels weren't enough. The engineers figured that if they could have multiple conversations on one channel, the system capacity could multiply several times. The digital cellular concept sends the telephone conversation in a data format. If someone were to monitor a digital cellular conversation, the sound would be similar to that of a fax machine when transmitting data.

Where Does the Phone Work? Where Can You Call?

By this point you should have enough cellular information to become dangerous. When the call travels from the mobile to the local telephone office, it is easy to understand how cellular phones can call anywhere a landline phone can call. The only restriction is the phone must be in a cellular service area. This has been a great marketing tool for cellular. I can recall a conversation in 1990 with a man who couldn't believe a cellular phone could call the Soviet Union. Isn't technology great?

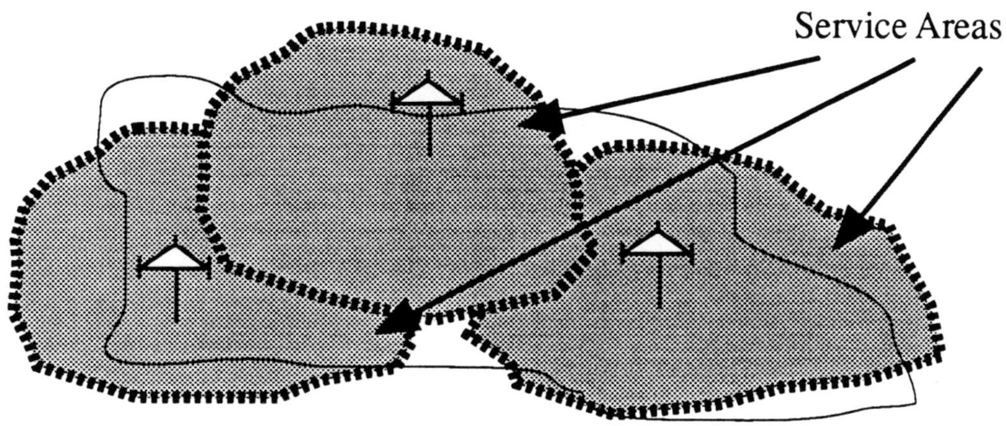

Cellular users often ask where their phone will operate. The answer is any area with an AMPS compatible cellular system, provided the phone has local service or the user is a welcome roamer. Roaming refers to using a phone in a non-home system. For more information on roaming, see Chapter 3, *Roaming*.

Tones and Messages

Cellular phones can produce a unique set of tones from the earpiece or speaker. The tones suggest a problem with the phone, location or system. The two common tones generally heard by cellular users are the fast busy and European siren. The fast busy emits when a user attempts a call that the network cannot process. Systems will typically send fast busies to suggest overcrowded cells or system problems.

The second type of tone is the European siren. Phones sound the siren to say the mobile can't talk to the system. If a cellular user attempts a call and the phone's No Service light is on, the siren will sound.

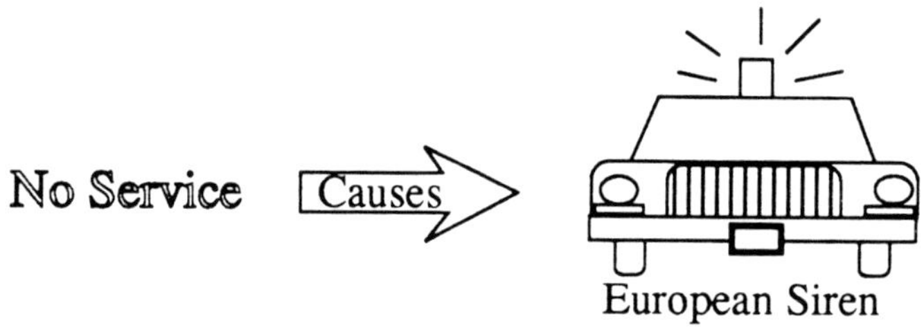

It is helpful in troubleshooting to remember the system sends fast busies and phones produce the siren.

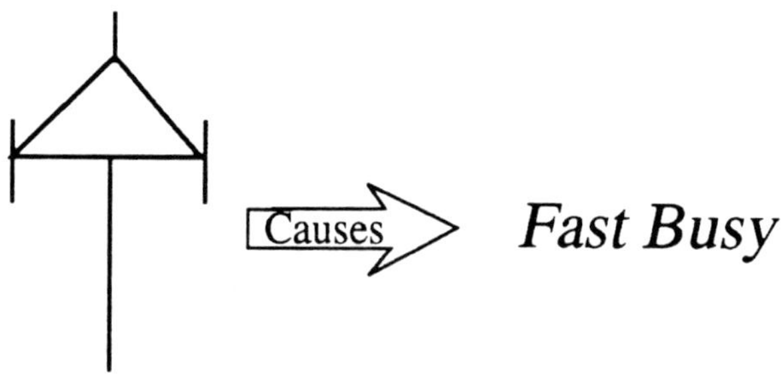

Besides tones, the network can notify the customer with pre-recorded messages. The messages vary in content, depending on the network. Consult your carrier for the current list of messages. A generic list of potential messages is listed on the following page. The messages tell the cellular user why the call was denied by the system. Several carriers also use a message number to help customer service departments quickly verify the customer's problem. A message might say "Your phone is not authorized to use our system, message 3". "Message 3" may mean that the ESN did not verify, causing the call to be blocked.

Sample Messages	**Possible Cause**
"Our records show your account is past due"	Customer owes money
"All circuits are busy, please try your call again later"	Busy system
"Welcome to our network. Your phone is not authorized to operate on our system. Please have your VISA card handy"	Roamer customer
"Your phone is not authorized. Please dial XXX for assistance"	Wrong ESN
"Security Alert! We need to talk to you"	Stolen phone

System Options

Cellular systems have options such as call waiting, conference calling, call forwarding and electronic message service. The network also can block incoming or outgoing calls if requested by the user. Some cellular phone models have self-contained answering machines. Fortunately, this feature never became popular. An advantage to a network operated message service is the customer doesn't pay airtime for any recorded messages. If a user depended on an answering machine inside the phone, airtime would be incurred for each message.

Another customer option is a tone pager. The pager beeps whenever the system's electronic answering machine has a new message for the cellular customer.

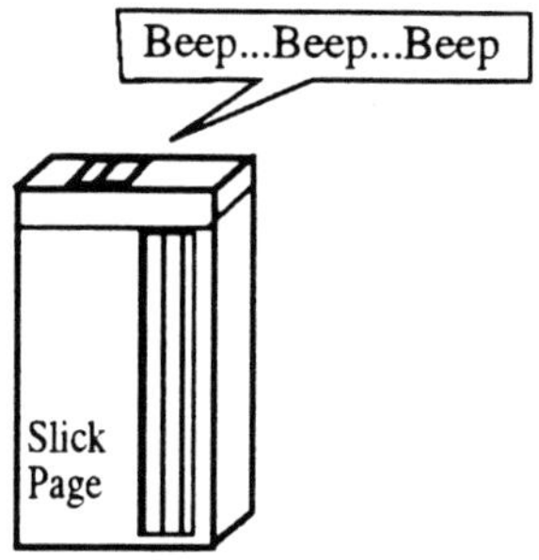

Customers initiate call waiting and conference calling in ways similar to landline phones. To initiate call waiting, the network beeps the customer during a call. The customer presses the SEND button and the call transfers to the second incoming call.

By pressing the SEND button a second time, the user returns to the first conversation. Conference calling begins in the same manner with the SEND key. Pressing the phone's SEND key during a conversation produces a tone called a hook flash. The hook flash is a fancy way to describe a short burst of tone from the mobile to the cell. If you were to listen to a hook flash, you would hear a high pitched (10 Khz) tone that lasts for 4/10 of a second. In other words, it's just a blip.

3 Roaming

What Is Roaming?

Cellular telephones are programmed to operate in a specific cellular system, called the home system. A phone operating in a non-home cellular system is called a roamer. Most phones have lights in the display to notify the cellular user if the phone is roaming.

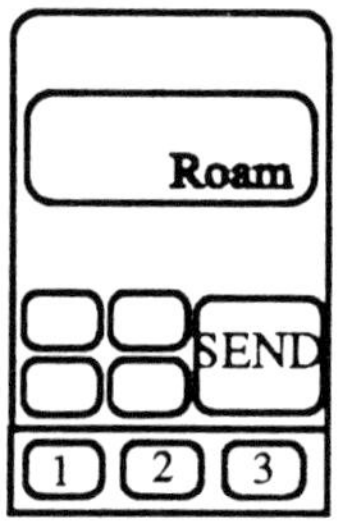

The Roam light generally illuminates solid if roaming in a home type system and flashes for non preferred systems. Each cellular system has a discrete system identification number. The cellular system control channels notify the phones, via data, of the system identification number. If the system's ID number is different than the phone's programmed ID number, the telephone's Roam light displays.

A Seattle customer from the B carrier drives to San Francisco. The technician programmed the phone to system 00006. This is the system ID for the B carrier in Seattle. The customer approaches San Francisco and notices the Roam light in the phone's display. This tells the customer that the phone is in a wireline cellular system, yet not the home wireline system. If the phone was in the home system, the Roam light would not display. San Francisco's system ID is

00040. The Roam light illuminates because the existing system ID (00040) doesn't match the programmed system ID (00006).

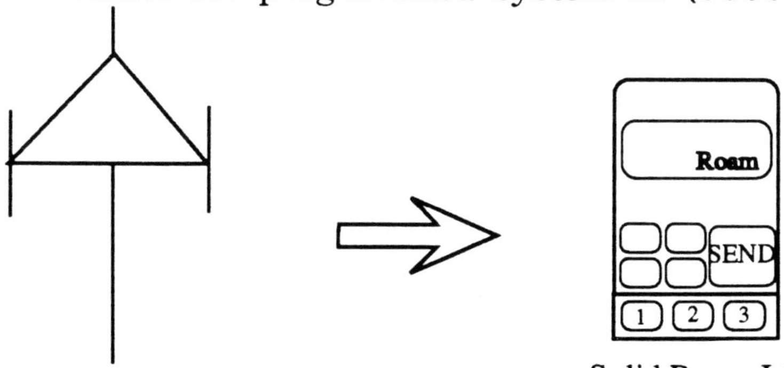

Non-Home "Like" Systems

Solid Roam Light

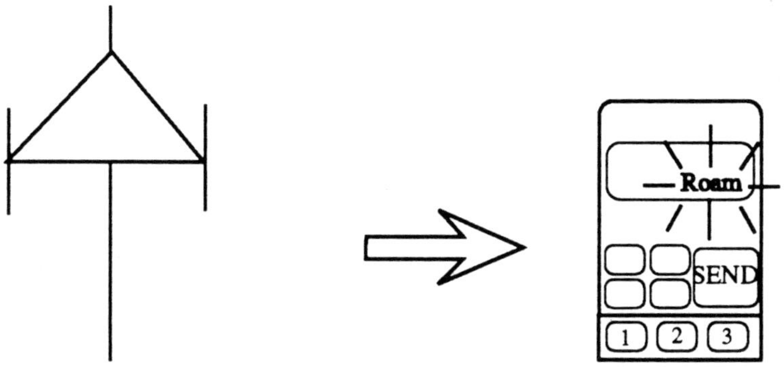

Non-Home "Unlike" Systems

Flashing Roam Light

An understanding of network operation, wireline versus non-wireline and system select are necessary to grasp roaming ideas. If you have any questions about network operation, refer to Chapter 2, *The Network*. For any questions about System Select, see Chapter 4, *Customer Equipment*.

Roamer Agreements

A customer who frequently roams will learn quickly the hoops and obstacles of roaming. Cellular carriers negotiate with other systems and create roamer agreements. A roamer agreement allows non-home customers to operate with specific carriers. The roaming agreements are reciprocal in that the allied customers are welcome to roam in other systems in the pact. This is where it starts to get confusing. Since the customer's home carrier doesn't have a roaming agreement with all cellular systems, the customer's phone won't be allowed to roam in all systems.

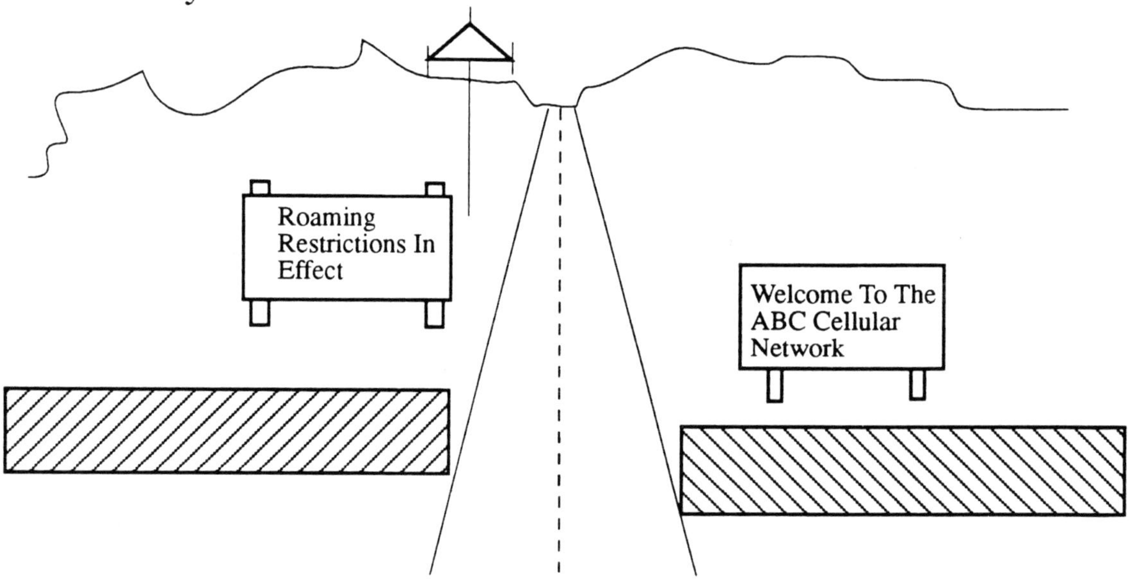

How does the customer know which systems and carriers have roamer agreements with the user's home system? Fortunately, most carriers provide customers with a list of cities where the phones will operate. A wireline carrier may have agreements with wireline and non-wireline carriers in other markets. The customer must know how to operate the phone's system select feature to access any non-preferred systems. System select allows the user to select the "A" carrier, "B" carrier, both carriers-"B" first, both carriers-"A" first or home only. Can you see the danger if the system select feature is set to "Home Only" and the customer drives to a non-home system? The phone's No Service light would never extinguish outside the home system.

Roaming is complicated enough to justify carrier "Roamer" departments and publishers spending countless hours collecting roaming data for user distribution. For a listing of roamer guides, see Appendix C, *Additional Resources*.

Simplified Roaming

Roaming, with its complexities and details, has been a confusing subject for years. Customers have a simple vision of how roaming should work. According to cellular customers, the phone should place and receive calls with no extra digits to dial. The cost of roaming should be the same as operating a phone in the home system.

Reality steps in, and again ruins a perfectly good dream. Some carriers have developed ingenious plans to forward calls to the user's system when roaming. Some even have a feature known as autonomous registration. With this feature, the network identifies a roamer customer when the phone enters the system.

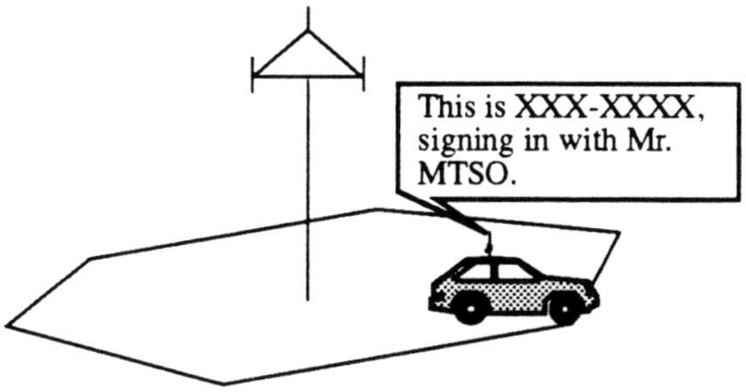

Autonomous Registration

The phone must be on to activate this feature. The network registers the mobile and can forward any incoming calls to the visiting cellular customer. Since roaming involves additional paperwork, it costs more than a phone operating in the telephone's home system. The job of producing a user-friendly roamer program should keep the carriers busy for several years.

Placing Calls in a Non-Home System

Sending and receiving calls in non-home systems can be tricky. The Seattle customer wishing to place a call while in Miami must dial the called party's 10-digit phone number. The area code is necessary even for roamers to call across the street. The Seattle customer is nearing the Miami airport and wishes to call the airlines for flight information. The 7-digit phone number and the local area code of 305 must both be entered. Frequent roamers will generally store 10-digit numbers in the phone's memory to avoid dialing delays when in non-home systems.

Roamer customers must dial the area code and seven digit telephone number to complete the call.

Placing a call in a system without a roamer agreement can be confusing to the cellular novice. The user may attempt a call and receive the frustrating message, "Welcome to the XYZ network. If you wish to place a call, have that credit card handy." An operator breaks in to establish the account and start the credit card debits. Networks charge cellular customers initiation fees, daily access charges and $X.XX per minute. It is important for the user to understand the roaming charges up front or the "sticker shock" can be painful.

Typical Roamer Charges

Account set up fee	$XX.XX
Daily fee	$XX.XX
Cost per minute	$XX.XX
Long distance	$XX.XX
Cost per visit	$XX.XX

Receiving calls in a Non-Home System

Receiving calls in a non-home cellular system can occur multiple ways, depending on the network. The most common method is the roamer access number. For example, you may wish to call a Portland customer who is sightseeing in Toledo, Ohio. You know the phone is on, operating within the boundaries of the Toledo system and set to the "B" (wireline) carrier. To place the call, first dial the 10-digit roamer access number for the wireline carrier in Toledo. A recording will instruct you to enter the 10 digit mobile number. Provided the customer is a welcome roamer on the Toledo system, the cellular phone should start ringing. As you can see by this scenario, there are a number of variables.

The goal of customers and carriers is hassle-free roaming. Another type of roaming involves customer registration in non-home systems. The user registers with the non-home network by entering a specific keypad sequence when entering the system or the network identifies foreign users with autonomous registration. Once the non-home system knows of the phone's existence, it notifies the home system. The user's home system forwards incoming calls to the phone operating in the non-home network. The cellular user is generally responsible for all long distance charges for the forwarded calls. An incoming caller doesn't know that the local system forwarded the call to a different state.

A second variation of roamer forwarding places the long distance burden on the incoming caller. Let's suppose a Seattle cellular customer roams in a California cellular system. An incoming caller dials the local Seattle cellular phone number. The cellular user is in California and has informed the home system to have all Seattle calls forwarded to California. The Seattle network informs incoming callers of their long distance liability. The calling person receives a recording that the user is operating in another system. The calling party hears the message, "You will be responsible for any long distance charges. Do you wish to complete the call?" If the caller agrees to pay the long distance, the home system connects the call to the non-home network.

Please understand that roaming methods vary, depending on the network, carriers and switch options. For specific network information refer to your cellular carrier roamer department.

When Do I Pay Long Distance?

Long distance charges for roamer service can be confusing. If a Los Angeles cellular customer is using a portable while skiing at Vail, Colorado, will there be long distance charges? Calls to the town of Vail will incur airtime and roamer charges only. A call back to Los Angeles from Vail also will be charged long distance. The phone is programmed with a Los Angeles number and local calls to Vail do not travel back to the Los Angeles cellular system. Vail calls, in this example, are not long distance connections.

Roamer Fraud

There is an ugly side to the roamer business. Unfortunately, some people have figured out how to beat the system with counterfeit phone numbers and ESNs. The phone bandits sneak onto the cellular systems and place countless calls to places with names that are difficult to pronounce. The airtime and long distance charges can't be collected and guess who absorbs the loss? The odds are high that paying cellular customers absorb the losses.

When a user places a phone call, the system verifies the phone number and electronic serial number with the carrier's data base. If the transmitted numbers agree with the network's data base, the call connects.

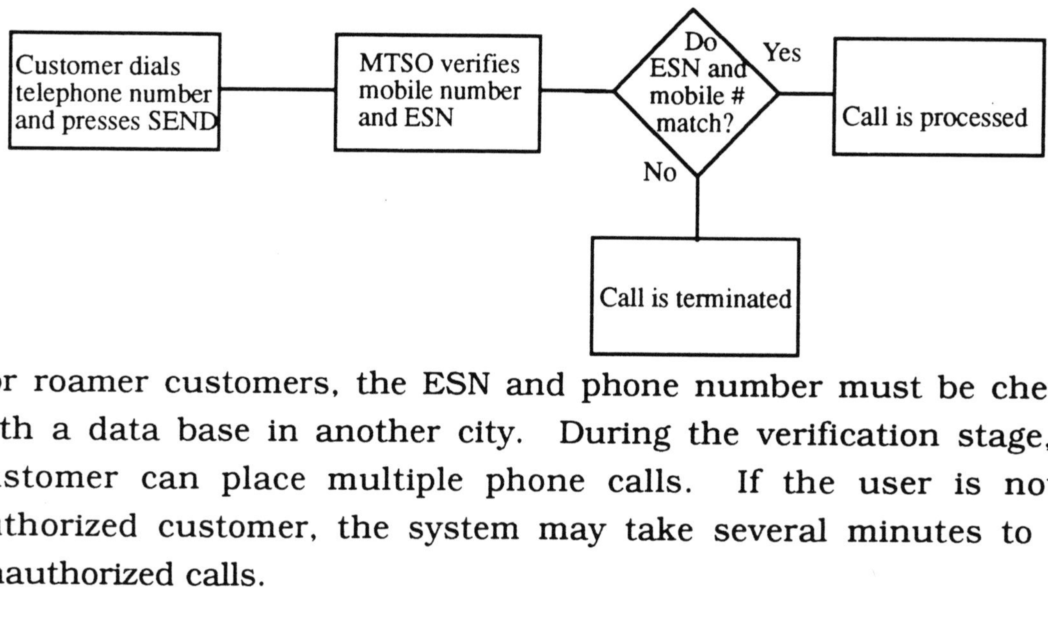

For roamer customers, the ESN and phone number must be checked with a data base in another city. During the verification stage, the customer can place multiple phone calls. If the user is not an authorized customer, the system may take several minutes to stop unauthorized calls.

To combat roamer fraud, several systems use autonomous registration to ask the mobile for its phone number and ESN. The system has this information before allowing the first call to be processed. The beauty of this system is its proactive design.

Keeping track of the phone numbers and ESNs is a labor intensive task. Clearinghouses collect this data and distribute the "go" and "no-go" lists to carriers who subscribe to their service. Loss of revenue from roamer fraud is a single digit followed by a string of zeros. Carriers and long distance providers hope to eliminate the problem soon.

Multiple NAM Phones

The term "NAM" stands for Number Assignment Module. This is the component inside the phone that stores the telephone number. A phone that can operate with multiple phone numbers is called a multiple NAM phone. To select a different phone number, the user enters a keypad sequence and the phone switches home systems. The multiple NAMs must be programmed initially to operate this feature.

A Colorado customer has an office in Denver and a second office in San Diego. Since the phone spends equal time in each city, the user wants one number for Denver and a second number for San Diego. This saves the customer roamer charges and simplifies incoming calls. The technician programs the phone with two telephone numbers. When the customer is in Denver, the first number is used. A simple keypad sequence entered upon arrival in San Diego, switches the phone to the second number. The customer pays number activation fees for a phone number in each city. Depending on the system usage, this can be an advantage to volume users in limited markets.

Inter-System Handoffs

As systems mature and new RSAs fill the coverage gaps, a new problem perplexes the cellular engineers. How does a customer operating in one system handoff to another adjoining system? Handoffs within a system are simple because a single MTSO controls multiple cells. The challenge is to handoff the phone to another MTSO. The most important element is politics. Cellular carriers create alliances with adjoining systems to provide "seamless" service for their customers. If adjoining carriers decide not to cooperate, the customer is the ultimate loser.

Part 2

Phones And Telephone Service

Customer Equipment

Telephone Activation

Troubleshooting

Equipment Repair

4 Customer Equipment

CELLULAR PHONE TYPES AND OPTIONS

Mobiles

In the mid 1980s mobile phones were the most popular type of cellular telephone. The current nationwide trend is shifting toward portable and transportable phones because of instant activations and no requirement for a permanent installation.

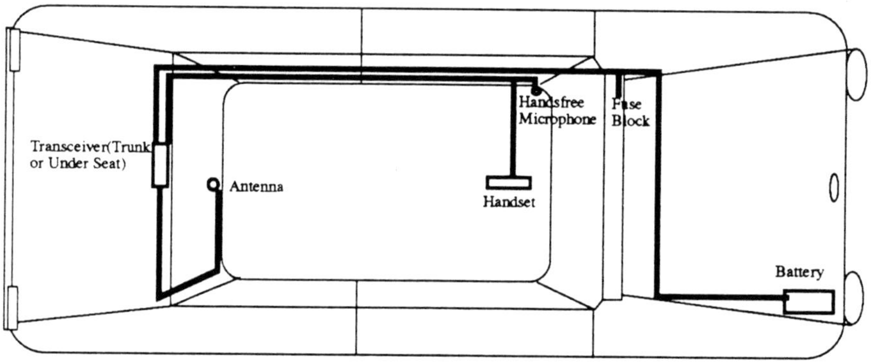

Typical Cellular Mobile Installation

The initial mobile installations involved expensive and elite automobiles. As the cellular idea gained speed, work trucks and less expensive vehicles began to dot the highways. It was common to see a vehicle equipped with a curly antenna and a driver who talked to the car's sun visor.

Most cellular customers were purchasing their first car phone. It is amazing to see new cellular owners who can't continue their business or social life if the phone leaves temporarily for service. I never saw a customer kick and scream like a 3-year-old, but it was close.

The cellular mobile phones have the maximum power output of 3.0 watts. This gives the mobile customer the best possible service, with the transportables and portables falling slightly behind. Besides the greater power output, mobile installations have less chance of becoming cellular torpedoes during an accident. Another advantage of mobile installations is the antenna mounts on the vehicle's exterior. An outside antenna gives the cellular user the best possible reception.

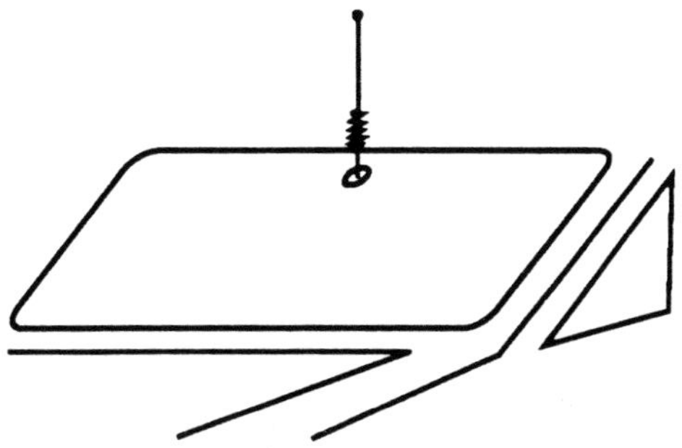

Outside Antennas Improve Transportable Reception

The one disadvantage of the mobile installation is a lack of portability. Mobile phones can generally be removed and placed in a transportable. Because this extra step takes time, few users want the additional work when leaving the vehicle. This may explain the growing customer support in portable and transportable phones.

Portables

The cellular portable has increased in popularity throughout the years. Portables are convenient, light-weight and operate in most environments. Improved system coverage, equipment size and portable prices have fueled the popularity. The first generation portables were expensive, with prices over $3,000.00. This is quite different than today's portable prices.

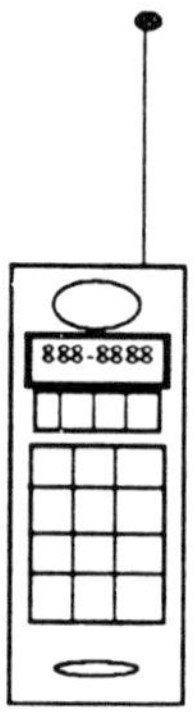

As the cellular systems mature, the portable has become a reliable and addictive communication tool. The early cellular systems had many "holes" in coverage, giving portable a lower grade of service than mobile telephones. The mobiles operate with a 3.0 watt output versus the 0.6 watt portable power output. A phone's power output is no problem unless the system coverage is weak or uneven. As networks add new cell sites and system integrity improves, portable customers benefit from the enhanced service. The portable has been popular for travelers because of the compact size and instant phone service in remote markets.

Cellular portables can be operated inside buildings, cars and other shielded environments with only one minor problem, call clarity. Cellular telephones are nothing more than a computer and a two-way radio in a shell. A radio's reception is dependent on location, antenna and power output. The computer aspect of the phone is dependent on data transmission between the phone and the system. The most volatile portable sales involve a sales technique known as "Over Promise, Under Deliver".

Customers should understand there are minor limitations to cellular communications, yet the benefits considerably outweigh the flaws. A realistic sales pitch describes what the phone can do and diplomatically explains the portable's limitations.

Another portable misunderstanding is that portables operate the same inside or outside the vehicle. The portable's lower power output and the vehicle's metal shielding decrease reception. The outside antennas used with mobile phones give vehicle-mounted telephones a noticeable reception advantage.

Portable Vehicle Adaptor Kits

Several manufacturers have developed vehicle adaptor kits allowing cellular portables to operate with the vehicle's power and use an exterior antenna. Portable vehicle kits vary in design and options. Some car kits have booster amplifiers to increase the portable's power output to that of a mobile phone, 3.0 watts. A common misunderstanding is that all vehicle kits automatically raise the portable's power output to 3.0 watts. Some early kits provided a power source and antenna function only.

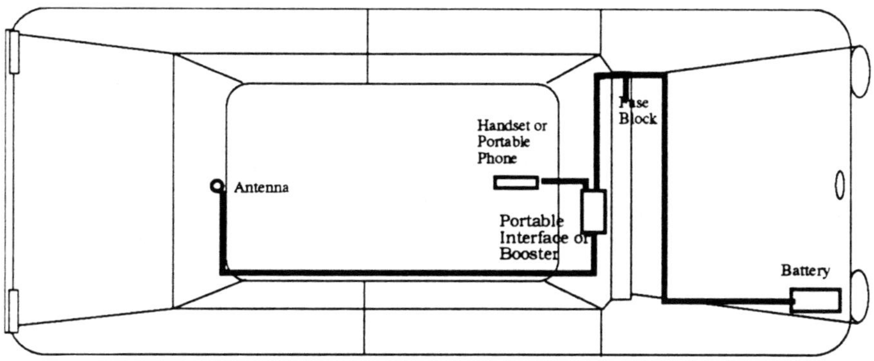

Typical Portable Mobile Kit

With a non-boosting kit, the portable's output power actually decreases from the additional cabling and antenna connectors. A vehicle adaptor kit will never increase the portable power beyond cellular's 3.0 watt maximum output.

Another type of vehicle adaptor kit is a portable cigarette lighter adaptor. This adaptor is also called a "battery saver." To use the adaptor, remove the portable's battery and slide the adaptor on the phone's battery mount.

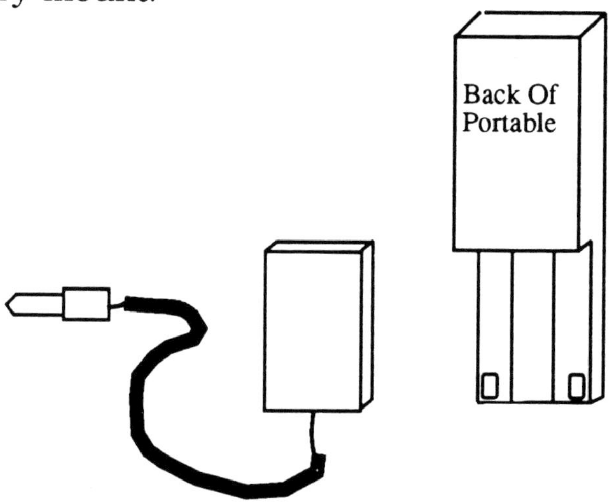

Portable Cigarette Lighter Adaptors

The adaptor then plugs into the vehicle's cigarette lighter. The adaptor uses the car's voltage system for power, thus eliminating the need for a portable battery while in the car. The phone's signal passes through the portable's antenna with this type of accessory. Battery eliminators are available from phone and after-market manufacturers.

Transportables

Transportables, like portables, are a self-contained unit with a transceiver, antenna and battery. The transportable, or transpack, design is somewhat cumbersome in comparison to that of the portable.

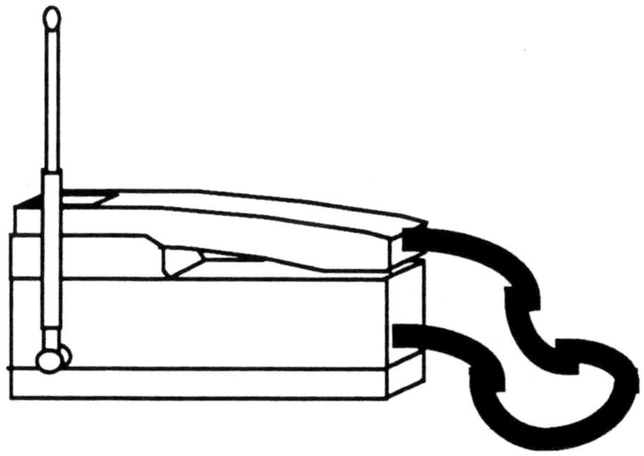

The main advantage of a transpack cellular phone is the power output of 1.2 to 3.0 watts, similar to the mobile phones. This results in better reception in a fringe area than the portable with 0.6 watts. Transportables also have a connector to attach an external antenna, a plus when operating the phone inside the vehicle. Most models also include a cigarette adaptor to use the vehicle's power for the phone, thus charging the phone's battery while in transit. This type of phone is popular in applications where the phone may be moved throughout the day.

As with the portable, transpacks operate best in the vehicle when used with an exterior antenna. Without an external antenna, the vehicle provides a blocking effect on the phone's signal. Transpacks operating in the vehicle without an external antenna have, on occasion, caused damage to vehicle "on-board" computers. Internal vehicle antennas cause excessive radio frequencies to bounce around the vehicle's passenger compartment. The best suggestion, when using a transpack, is to use an exterior antenna to maximize performance.

Transpack Car Kits

Vehicle kits, also called second install kits, are available for most transpacks. With the vehicle kit, the user attaches the transceiver and handset to the data cabling installed in the vehicle. If the customer wishes to move the phone from transpack operation to the car kit, the handset and transceiver are simply connected to the car. When installing a second install kit, the transceiver is generally mounted in a convenient location for easy installation and removal.

Cellular Phone Batteries

Transpacks usually have lead acid (gel cell) or nickel cadmium batteries. Portable cellular phones use nickel cadmium batteries. It is vital that the customer is informed of proper battery care and operation at the time of delivery. The best advice is to take a few moments to read the battery instructions included with the telephone. Since battery abuse can cause irreversible battery damage, some manufacturers have shortened the battery warranties to 10 days. Others include battery care instructions on the first page of the user's guide to emphasize proper battery care and handling.

Lead Acid-Gel Cell Batteries

The gel cell battery is similar to a car battery in charging characteristics. Gel cell batteries love to be charged. The phone can be used for ten minutes, placed on the charger for a few minutes then placed back into service.

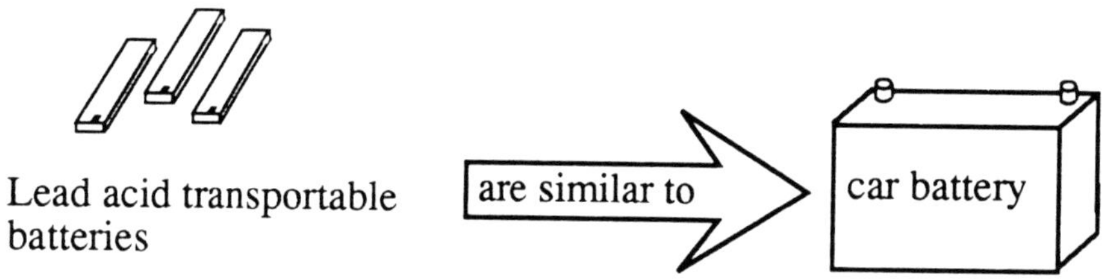

This is the same treatment a car battery receives when you start the car. This starts the battery charging cycle. If you stop at the store, the battery will rest. Then when you start the car again to drive home, the battery is back in the charge cycle. Throughout the day the battery is charging, resting or discharging. Lead acid batteries are designed for this type of service.

The quickest way to ruin a car battery is a process known as deep discharge. For example, it is foolish to leave your car's dome light on all night while the car sits in the garage. The following morning your battery has a major hangover. The car won't start because the dome light drained the battery. The car can be jump started, but each deep discharge causes non-reversible damage to the battery. The same care must be observed with lead acid cellular telephone batteries. Do NOT deep discharge a gel cell battery. Battery abuse is usually not covered by a manufacturer's warranty.

Nickel Cadmium Batteries

According to several studies, nickel cadmium or ni-cad batteries operate best when full cycled. Full cycle means to charge the battery completely and not place the battery on the charger until fully discharged. This full cycle characteristic establishes a memory in the ni-cad battery providing maximum performance.

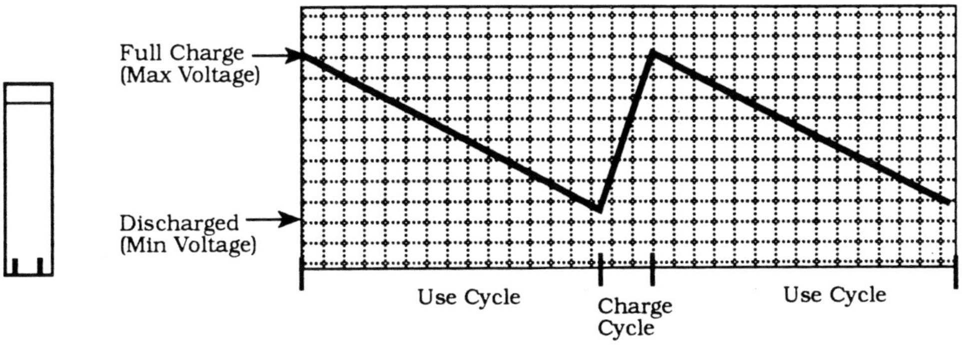

Charging Ni-Cad Batteries

Some cellular manufacturers claim their batteries do NOT have memory. The best advice is to read the battery instructions included with the portable or transpack.

A suspected defective nickel cadmium battery can be tested by most cellular repair facilities. The testing process, called cycling, involves a battery analyzer. This process charges the battery to capacity, then the battery slowly discharges to be measured for storage capability. The analyzer measures the battery with a standard known as an amp hour rating.

A battery with a 2 amp hour rating can be drained at a rate of 1 amp for 2 hours. The same battery can be discharged at a rate of 2 amps for 1 hour. Amps times hours equals amp hours. The point is that the battery analyzer cycles the battery and determines the battery's amp hour rating. If the customer complains of poor battery life, the test results are a scientific way of examining the battery characteristics. With the test results, a technician determines if the battery can be rescued.

Stationary Cellular Installations

Emergency cellular telephones have appeared in several markets alongside freeways. This application, sometimes called a "Call Box", is popular because of the ease of installation and minimum maintenance.

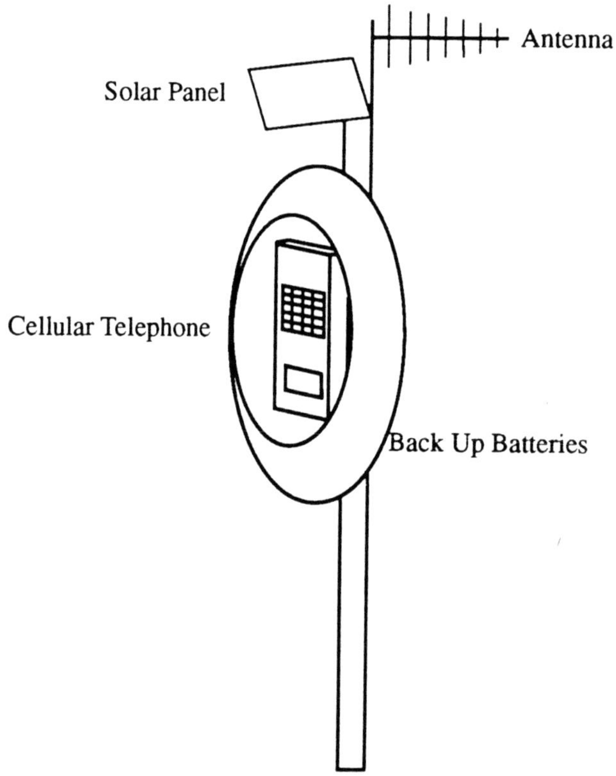

Roadside Cellular

The stationary phones allow a stalled motorist or accident victim to lift the phone's receiver at the roadside telephone and connect to a crisis switchboard. Most freeway phones are in areas with limited landline telephone service. The roadside phones receive power from batteries in the telephone booth. To eliminate the need for external battery charging, the phone booths have solar panels to charge the phone's storage battery. An external antenna gives the phone maximum call clarity.

Laptop Computers and Facsimile Machines

The day of the portable office is finally here! Several models of portable facsimile machines and laptop computers are available. The computers and facsimile machines connect to cellular phones with custom connectors or by using an accessory known as an acoustic coupler.

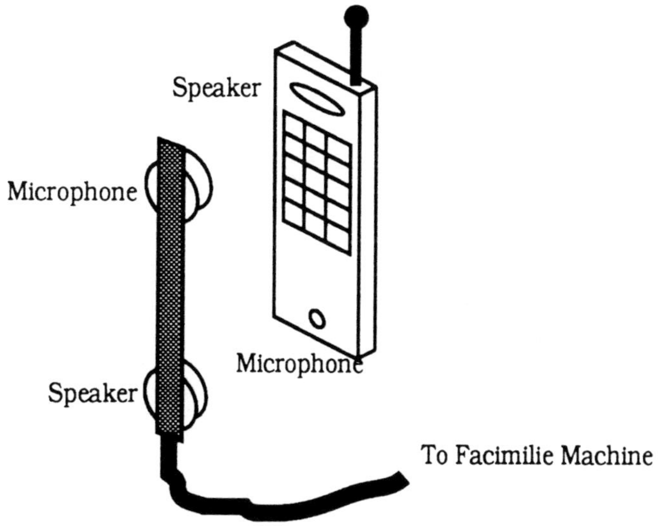

Acoustic Coupler

The acoustic coupler is a microphone and a speaker built into a single housing that can be pressed against the customer's cellular handset or portable. The coupler's microphone and speaker align with the phone's handset to talk and listen via the phone's handset. To start a facsimile machine or laptop computer, you must first originate the call from the phone's handset like a standard cellular phone call. The accessory provides a link between the accessory and the cellular telephone.

Facsimile machines also can be direct coupled with a RJ 11 accessory. The term RJ 11 is a fancy way of saying a home phone jack. If you unplug your phone jack from the wall, you will notice a female receptacle with 4 contacts. This standard is used in the cellular industry for connecting a facsimile machine to a cellular phone.

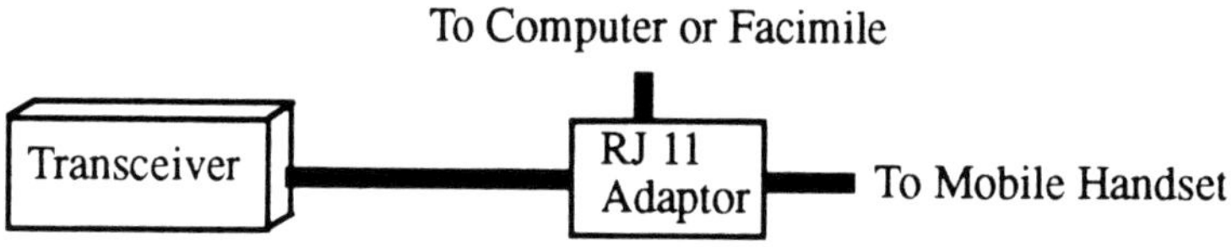

Some cellular telephone manufacturers offer RJ 11 adaptors for their phones to connect easily to computers and facsimile machines. There are also after-market manufacturers that produce RJ 11 devices for cellular phones.

One thing to remember when using a cellular phone to send data is the data rate, also called baud rate. This rate may be slightly slower than landline transmission to compensate for any irregularities caused by the telephone's signal. Make sure the modem operates at a rate compatible with cellular. If the rate is too fast, much of the data may be lost by signal irregularities. System handoffs during data transmission also can be a source of data problems. As the call transfers to another channel, a portion of the data may vanish.

Cellular Pay Phones

Few cellular manufacturers produce cellular pay phones. The market for this product has limited applications. Cellular pay phones operate in areas where a pay phone is needed yet no landline connections are available.

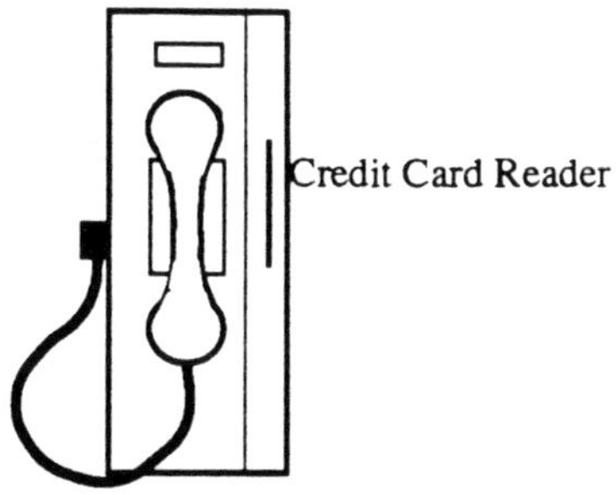

Cellular Pay Phone

Examples include islands, tour boats, buses and rest areas. Some cellular pay phones look like standard credit card phones found in airports and hotels. The customer simply passes the credit card through the card reader and dials the desired phone number. The airtime and long distance charges are billed to the customer's credit card. Could you imagine the consequences if coin-operated cellular pay phones became common?

Cellular Coin Phone

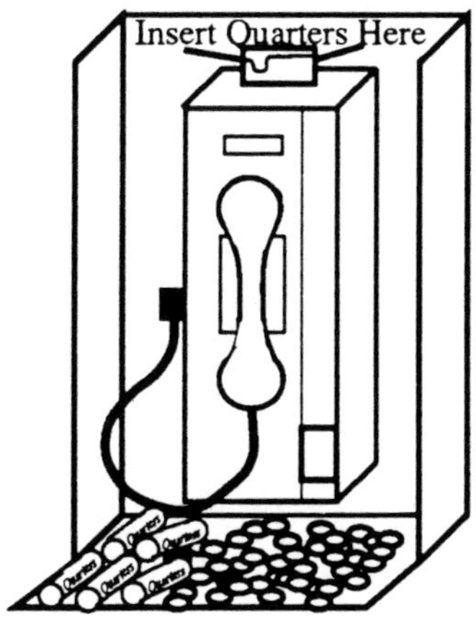

I can picture patrons loading $10.00 rolls of quarters to cover call costs and an attendant emptying the phone every hour, on the hour. Not a pretty sight.

Cellular Headsets

Cellular headsets can be added to most cellular telephone models. With the headset, the user's hands are free to operate a vehicle or other machinery. Another advantage of headsets is the audio quality. Standard handsfree options offered by most manufacturers place the microphone and speaker at least 12 inches away from the vehicle driver. As the microphone moves further from the source, the audio quality diminishes. The headset offers the advantages of handsfree with improved audio.

TELEPHONE INDICATOR LIGHTS AND CONTROLS

Cellular telephones have oodles of lights, knobs and switches. This section explains the lights and controls common to most cellular models. Since there are several hundred models of telephones, it would be futile to be model specific. With this overview you should have the knowledge to operate the basic features for most models. A glossary of *Phone Features* is included in Appendix A to provide additional information on other cellular phone features. The individual phone user's guides have model specific feature information.

Roam Light

The Roam light notifies the cellular user if the phone is in a non-home system. The light also will illuminate if the telephone jumps to the competitive system in the home area.

The light will generally display solid if on a like system and blink for non-preferred systems. In other words, a phone programmed to system ID 00006 (Seattle B Carrier), would show a blinking Roam light if operating on system ID 00047 (Seattle A Carrier).

The Roam light displays when the phone's programmed system ID doesn't match the system ID sent on the network's control channel. A phone mis-programmed with the wrong system ID will display the Roam light. If a phone has been programmed and immediately shows Roam, there is a problem. The phone must be reprogrammed or the user must check the phone's system select feature.

Some models show the Roam light in slow blink or fast blink configurations. The blinking speed tells the user which system the phone is listening to. A slow blink means home type roam and a rapid blink suggests a non-home roam. A phone programmed for the Atlanta A carrier would be a non-preferred roamer if operating with the B carrier in New York. The blinking Roam light is a common display method for a portable phone.

No Service Light

When a phone powers "on", it does a quick self-check and looks for the cellular system's closest control channel. During the looking stage, the phone's No Service light stays illuminated.

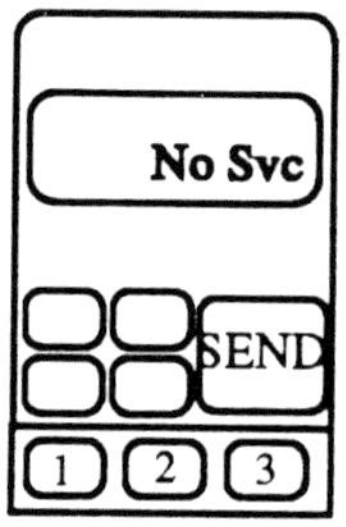

Provided the telephone is operating within a cellular coverage area, the No Service light should turn off within a few seconds. The light tells the cellular user that calls cannot be placed or received.

A phone leaving the cellular system also displays the No Service light. When driving away from a system, the light may illuminate intermittently and finally display solid. The signal from the network doesn't end at a specific point and you may find areas where the No Service light is off, yet you can't receive phone calls. This is because the phone hears the system, yet the phone's location is marginal, thus making it impossible for the mobile phone to respond.

In Use Light

The In Use light shows the cellular user that the phone is processing a call.

Entering a number and pressing the SEND button illuminates the light. The user presses the END button to complete the call and the In Use light extinguishes.

Signal Strength Display

A handy feature of some phones is the "signal strength display." Since cellular coverage varies throughout the system, this system's signal strength is helpful to the cellular user. The phone's display shows the system's signal as a series of "blocks" or "bars."

If the five bars display, the user knows the call integrity should be acceptable. In a situation where only two bars show, the cellular customer may wish to move to a stronger system coverage area. By having this information before placing the call, the chances of an inferior call decrease.

Volume Adjustments

Cellular telephones have three or four volume adjustments, depending on the model. The user adjustable volume controls are handset (earpiece), ringer, speaker (keypad tones) and an optional handsfree adjustment. It is common for phones to control multiple volume adjustments with one button.

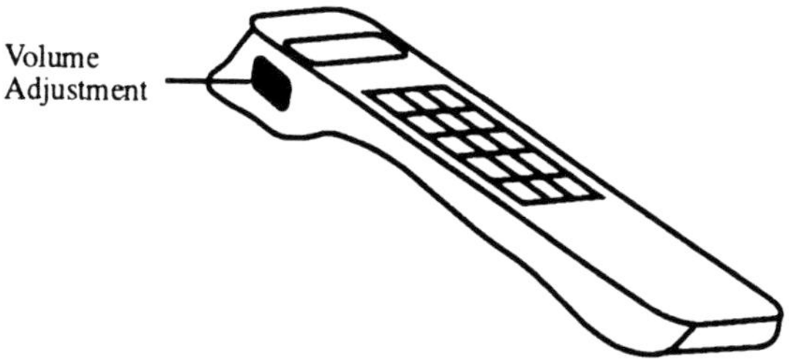

One example of this type of volume adjustment is MOTOROLA phones. Earpiece volume varies by pressing the volume button while in a conversation. Ringer volume changes by pressing the same button, provided the handset is out of the hang-up cup, with no call in progress. The same button adjusts the speaker volume when the phone is in the hang-up cup. Volume adjustment instructions can be found in the individual phone user's guide. Volume adjustments often confuse customers because of the multiple combinations with the same key.

System Select

In each market, there are generally two competing cellular carriers, the A and B networks. The key word is "competing". A telephone programmed to operate on A system is not usually welcome on the B system within the home market. This principle should not be confused with roaming, a situation where a phone roams on a system outside the coverage of the customer's home system. The point is, each market has two systems, and each phone is programmed to one system and is not welcome on the second.

To keep the phone on the proper system, the technician programs the phone at the time of purchase. An additional step, called system select, ensures that the phone stays on the designated home system. System select is accessible from the keypad and allows the following user choices.

Setting	Phone Effect
AB	Phone scans A channels. If A service cannot be established, the B channels are scanned.
BA	Phone scans B channels. If B service cannot be established, the A channels are scanned.
A Only	Phone scans A channels only.
B Only	Phone scans B channels only.
Home	Phone will operate in home system only. No roaming

Lock Key

The lock function prohibits unauthorized persons from using a cellular telephone. The phone is generally "locked" by pressing the LOCK button.

Calls incoming or outgoing will not be allowed with this feature activated. To unlock the telephone, the user enters a 3- or 4-digit unlock code that the technician programmed when the user purchased the phone. Most phones allow 911 calls even when locked. Some phones require the unlock code and an additional keypad sequence to unlock the phone.

Call Timers

Call timers can be useful in monitoring airtime use. The three common timers are individual, cumulative and resettable. Individual timers display the duration of the current call.

Cumulative call timers are similar to a water meter because they can't be reset. The displayed minutes or hours continue to increment to the timer's limit. Once the phone reaches the maximum digit limit, the timer starts again at zero. Cumulative timers often reset at 999 or 9999.

The third timer, the resettable, has a useful function. If a customer wishes to monitor the approximate airtime used during a billing cycle, the timer can be reset on day one of the new cycle. To reset the timer, the cellular user enters a code from the keypad and the timer returns to zero.

Timer Type	Function
Individual	Shows duration of last call
Resettable	Can be reset by user (i.e., monthly)
Cumulative	Can't be reset by user

Timer totals and airtime bill totals will never be the same. Since carriers use different increments and billing methods, phone timers can't be market specific. The totals displayed by the telephone should be used as an approximation of the airtime used.

Clear Key

When you lift the handset of a landline phone, a dial tone sounds. Cellular telephones are different. The user enters the desired telephone number from the keypad. Airtime starts once the customer presses the SEND key.

Cellular carriers want the customers to use as little time as possible to originate the call, thus saving carrier airtime. To ensure accuracy, the customer can view the dialed number before placing the call. If the number is wrong, the customer presses the CLEAR key to remove single digits or the entire dialed number.

Store and Recall Keys

Frequently used phone numbers can be easily stored in the phone's memory locations and recalled for speed dialing.

To store a telephone number, the user inputs the number from the keypad, plus STORE, and finally, a storage location. The number can be instantly dialed by pressing RECALL, the memory location and then SEND. The actual sequence and number of memory locations may vary, depending on the phone model. This feature is quicker than dialing a number manually.

Function Key

Could you imagine a cellular telephone with 35 keys? The phone manufacturers play an aggressive game of beating the competition in the features race.

The FUNCTION key makes one keypad button to do multiple functions. On one phone model the 4 button may show a "4" on the display when depressed. By entering the sequence FUNCTION plus 4, the call timer may display. A function key is an intelligent alternative to 35 keys. Depending on the model, function keys can access several features such as horn alert, system select or volume adjustments.

Mute Key

Picture this scenario. You are talking to your boss on your cellular telephone. He thinks you are on a sales call at the north end of town. You are in the south end of town by the airport and a jet is approaching the runway at 1,000 feet. To simplify your story, wouldn't it be handy to put your phone on hold? That is exactly what the MUTE key does.

The MUTE key places the call on hold, yet you can still listen to the incoming audio. This is an important feature for today's business person.

DTMF Tones

DTMF is a telephone industry acronym. The letters stand for Dual Tone Multi Frequency. I was recently conducting a workshop and one participant thought it stood for Don't Touch My Fone. I questioned that answer.

DTMF tones sound whenever you press a number key on a telephone. The tones can activate answering machines, remote control devices and answering services. If the technician accidently disables the DTMF tones during phone programming, the phone will not send tones during a call. This makes customers very frustrated.

DTMF can be sent during a call either by entering them manually from the keypad or by recalling the tones from the phone's memory locations. If the user dials a long distance number and is using a calling card, the memory location recall saves time. The 800 number for the long distance can be stored in location #1, the number dialed in location #2 and the customer's personal code in location 3. To dial the number, the customer recalls #1 and presses SEND. After receiving a tone from the long distance carrier, the user recalls location #2 and presses the SEND key a second time. Another series of tones sounds when the contents from location #3 are recalled and sent.

When you consider that the three memory locations held over 20 digits, you have again pegged the efficiency meter. Refer to the individual phone user's guide for DTMF memory dial. Some phones require that the user press the FUNCTION or RECALL buttons to access this feature.

Audible Call Timer

The audible call timer alerts cellular users of the airtime used. On most models with audible call timers, a tone sounds before each minute. If the network bills the cellular user by the minute, the tone allows the call to be ended and eliminates the charge for another minute.

Audible Call Timers "Beep" to Indicate Call Duration

This feature can be enabled or disabled from the keypad on most phones. Some phones require reprogramming by a technician to activate the audible call timer. If in doubt, read the phone's manual to figure out if the phone has an audible call timer as standard equipment.

THE CELLULAR INSTALLATION PROCESS

Vehicle Check In

The cellular installation employees greet the customer with smiles, handshakes and warm fuzzies. The paperwork, known as the Installation Instructions, contain the known transaction information, (i.e., vehicle make, model, year and phone type). The vehicle trunk and passenger compartment should be empty, and the customer agrees to a pick-up time.

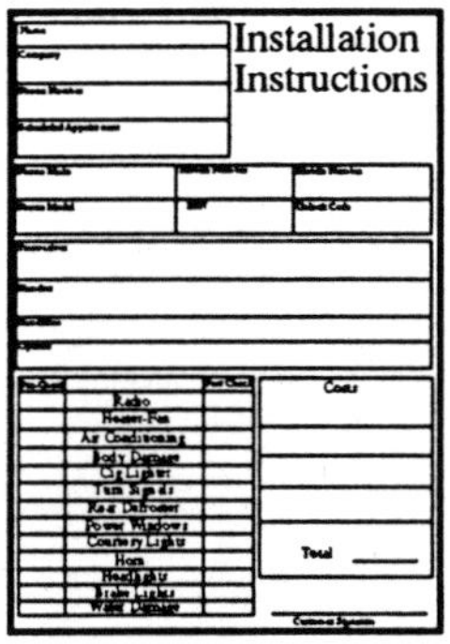

The customer shouldn't arrive with too many preconceived ideas of where each component will be located. The Installation Technician can generally suggest the best component locations. Component locations require planning to permit maximum accessibility and safety. Handsets and transceivers are methodically placed to allow access to other vehicle accessories. The vehicle driver and passenger safety is of primary concern when selecting component locations.

A review of the vehicle accessories checks for proper operation before starting the installation. Any existing vehicle damage (i.e., headliner flaws or body damage) is documented.

Pre-Check		Post Check
	Radio	
	Heater-Fan	
	Air Conditioning	
	Body Damage	
	Cig Lighter	
	Turn Signals	
	Rear Defroster	
	Power Windows	
	Courtesy Lights	
	Horn	
	Headlights	
	Brake Lights	
	Water Damage	

It is assumed the technician programmed and activated the cellular telephone before the starting the installation. For additional information on programming and activation see Chapter 5, *Telephone Activation.*

Antenna Installation

The customer selects an antenna at the time of purchase. Roof and glass mount antennas are the most common. Additional antenna types and characteristics are defined in *Cellular Antennas*, (Page 67). One main difference between the glass and roof mount antennas is that the glass antenna is a "no holes" installation.

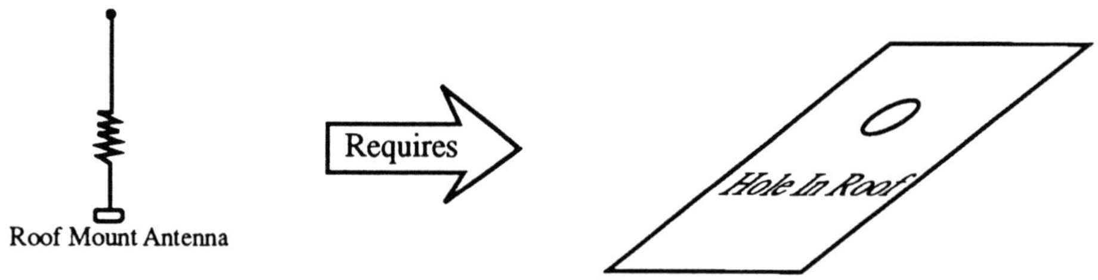

Roof mount antennas generally require a 3/4" hole in the vehicle's roof.

Glass Antenna

The glass antenna's four components are the antenna mast, exterior base, inside coupler and cabling. The antenna base mounts to the car's window, on the outside with double stick tape. The antenna mast inserts into the base on the vehicle's exterior.

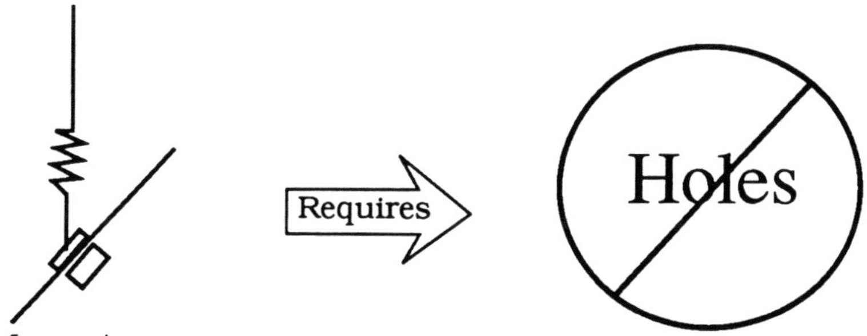

Glass Mount Antennas

The coupling assembly mounts to the window on the inside. The coupling box transmits the phone's signal through the glass. The final component is the antenna cable which carries the phone signal from the phone's transceiver to the antenna.

Roof Antenna

To mount a roof antenna, the installer drills a hole in the vehicle's roof. The antenna's mounting base secures to the vehicle's body and the antenna mast fastens with a retaining nut. As with the glass antenna, the antenna cable must be routed from the antenna to the transceiver. The antenna cable travels behind vehicle trim panels and under carpeting enroute to the transceiver.

Antenna Connector

Once the antenna cable reaches the transceiver, the technician installs a crucial component, the antenna connector. This tiny silver connector, if improperly installed, will cause unlimited customer complaints.

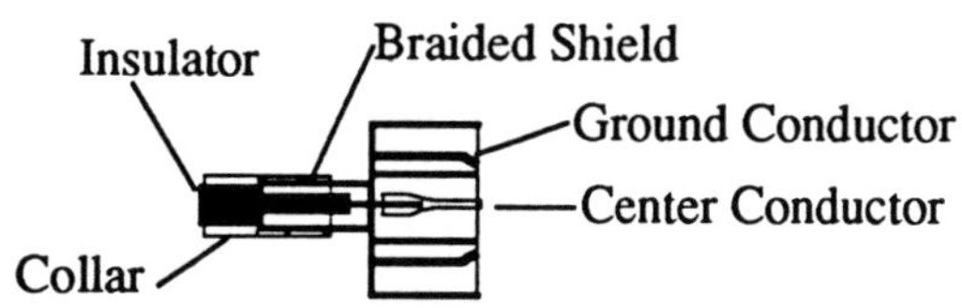

TNC Antenna Connector

The braided antenna cable is stripped back, exposing the conductor. The connector attaches to the antenna cable with a special crimping tool to insure a solid connection. With the connector installed, the technician secures the connector to the transceiver's antenna port.

Transceiver Installation

The technician selects a transceiver location in the vehicle's trunk or under a seat. The installer asks several questions before selecting the transceiver location. Does the trunk seal leak water? Is the trunk wet? Can the spare tire be removed with the transceiver mounted? Does the transceiver interfere with the trunk hinges? Is the area properly ventilated to prevent heat "build-up"? If required, can the transceiver be easily removed for service? Is the catalytic converter close enough to cause heat damage? Failure to ask any of the above questions may cause a free transceiver relocation or replacement.

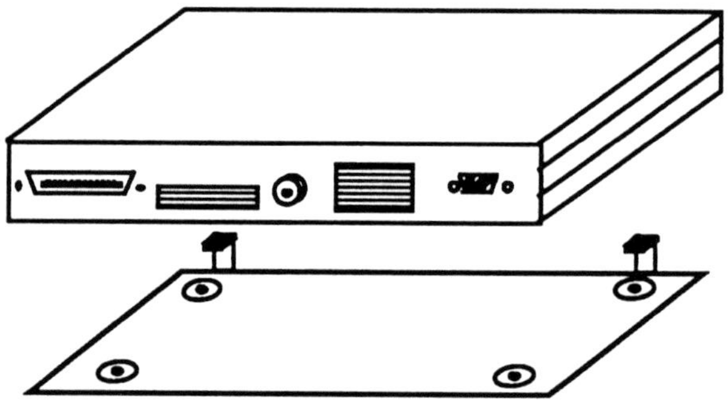

Transceiver With Bracket

With the best location selected, the transceiver mounts with screws to the vehicle body. The installer is extremely careful during this step to avoid drilling holes or driving screws into gas tanks, wire harnesses or brake lines. Blind drilling can be a very expensive lesson for an installer. Most installers have a blind drilling story that cost several dollars to correct.

Cable Routing

The phone's power and data cables, sometimes the same cable, must be routed from the transceiver to the handset and vehicle power source. The power cable generally contains 4 wires. These include a 12-volt constant source (battery), ignition sense (accessory), horn alert and ground. The ground secures to solid vehicle ground by the transceiver. The additional electrical leads connect to vehicle power at the fuse block or battery.

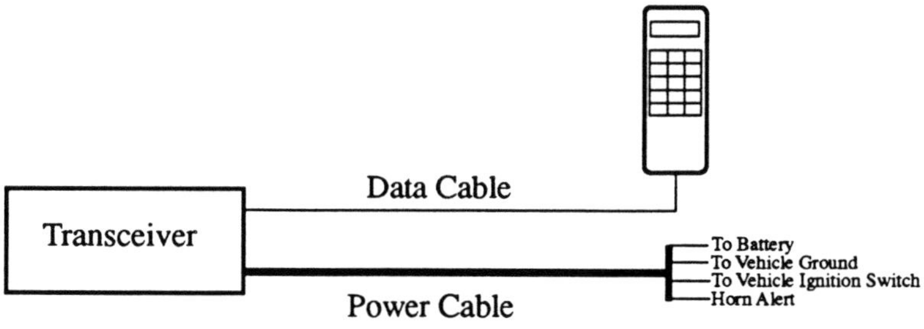

Cellular Mobile Cabling

The power leads connect to fuses for protection against a "shorted" wire. A fuse can be viewed as a safety valve. If a bare wire accidently contacts the car body, excessive current travels from the car's battery and through the phone's power lead. A fuse permits a maximum current of X number of amps, depending on the fuse rating. If a "short" occurs and the fuse rating is exceeded, the fuse disconnects the circuit.

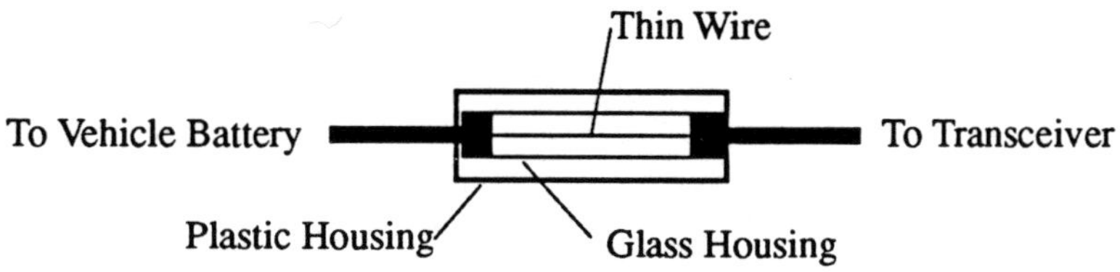

Fuse Components

The power and data cables route through the vehicle behind trim sections and under carpeting. The data cable connects to the telephone's handset and the power cable is attached to the vehicle's wiring.

Handsfree Microphone

The handsfree microphone mounts in a location close to the driver. Most microphones are directional, meaning location and angle must be considered when installing. Directional microphones hear noises directly in front of the microphone, yet not those sounds from the side.

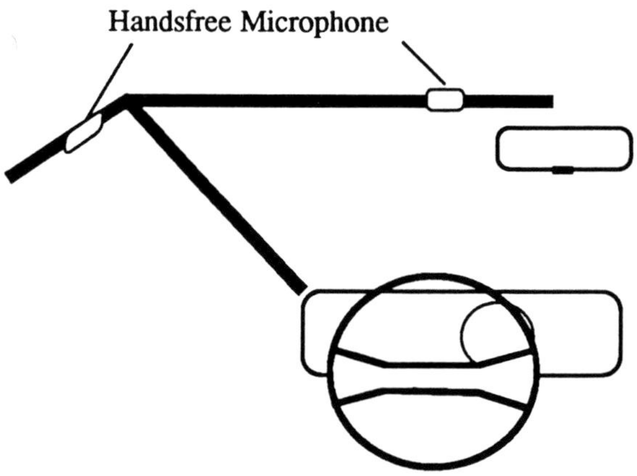

Recommended Microphone Locations

A popular handsfree location is on the headliner trim in front of the driver. Another standard microphone location is the driver's pillar. The microphone cable hides behind the windshield trim, inside the vehicle, and routes to the data cable or cradle for connection.

Handset Installation

Handset mounting requires similar questioning as with the transceiver. Is the location safe for the driver and passengers? Can the driver easily see and reach the handset? Once a location is selected, the handset mounting post or "clamshell" mounts to the vehicle with screws.

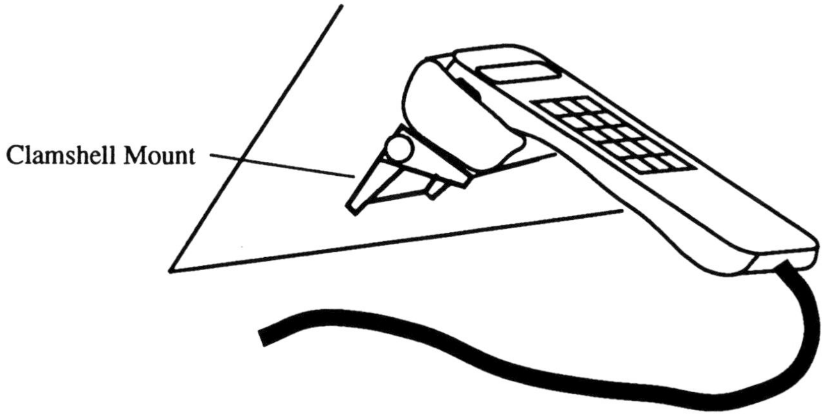

A clamshell is a style of handset mounting hardware that opens like, you guessed it, a clam. The clamshell fastens to the base of the phone's hangup cup.

CELLULAR ANTENNAS

Glass Antenna

Cellular antennas are available in several shapes and sizes. The most common cellular antenna is the glass mount. The greatest advantage of the glass mount is the "No Holes" installation. The antenna has four components: the coupling box, mast, antenna base and cabling. The antenna's coupling box mounts on the glass, inside the vehicle. The antenna and mounting base install on the car's exterior. The signal travels through the glass from the antenna base to the coupling box.

Most glass mount antennas are preset by the factory for maximum performance and require no adjustments in the field. An adequate antenna mounting location allows the antenna to extend above the vehicle's roof-line.

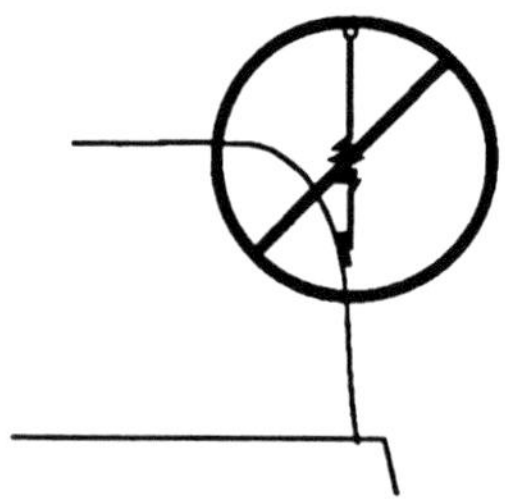

Antennas Should Extend Beyond The Vehicle Roof

This is particularly critical when selecting a location on a pickup truck rear window because the top of the window is several inches below the vehicle's roof line. If the antenna mounts below the roof line, the phone has substandard performance.

Each glass antenna should be checked with a wattmeter upon completion of the installation. A wattmeter connects in line with the antenna cable to measure the antenna signal.

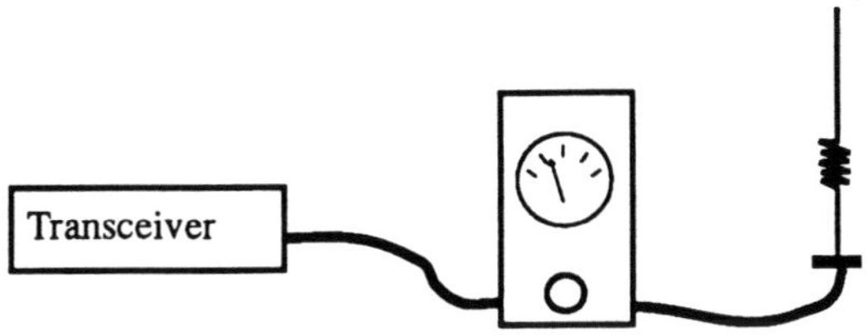

Testing System With Wattmeter

The wattmeter pin-points any cabling, connector or coupling problems at the time of installation. Cabling problems can occur if the antenna cable is pinched, pierced or has a manufacturing defect. A shorted or improperly crimped antenna connector will be apparent from a wattmeter test. If the technician suspects a window with high metallic content, the glass is tested with a capacitance tester to verify compatibility. The capacitance tester also exposes problems with window tinting, defroster wires and leaded glass.

Roof Antenna

The roof mount antenna is another common cellular antenna. As the name implies, this antenna installs on the vehicle's roof.

Roof Antenna

This antenna should NOT be mounted on the cowl, trunk or other surface below the vehicle's roof. The roof mount antenna requires a ground plane, thus making "L" bracketed installations substandard. A ground plane refers to a flat metal surface.

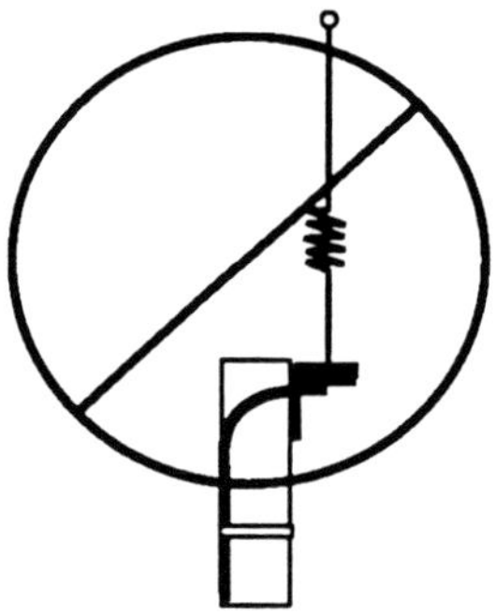

Improperly Installed Roof Mount Antenna

For cellular applications the metal surface must be at least 12 inches in diameter. Thus, the vehicle roof is the only choice.

Technicians inform customers of the 3/4-inch mounting hole required for a roof mount antenna before starting the installation. Failure to inform before drilling is another expensive lesson in customer service.

Cellular roof antennas should be removed before entering a car wash. I met one customer who forgot to remove the antenna. It stretched to 15 inches in length. He called it a high-gain antenna. I called it garbage.

Elevated Feed

The elevated feed antenna is another cellular antenna style. This antenna requires no ground plane and is several inches taller than the glass mount or roof mount antennas.

Elevated Feed Antenna

Elevated feed applications include trunk mount, cowl mount, installation on non-metal surfaces and instances where a ground plane is not available. They are also found "L" bracketed on truck lumber racks and mirrors or mounted or on motor home air conditioning housing.

Temporary Antenna

Temporary cellular antennas have many cellular applications. Models include magnetic mount, window suction and window clip antennas. As the name implies, cellular users can easily install and remove temporary antennas.

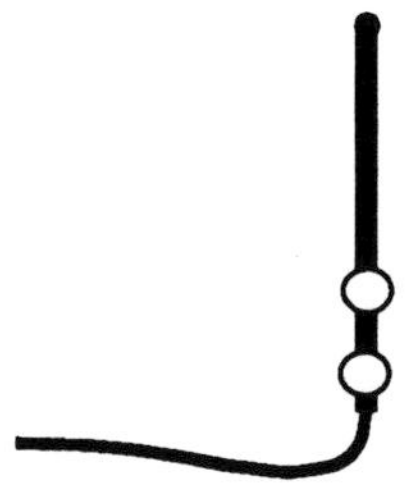

Suction Antenna

Temporary antennas are helpful for a customer operating a transpack inside the vehicle. Without the external antenna, the vehicle shields the transportable's signal, causing inferior reception.

One hazard with magnetic mount antennas is their tendency to scratch the vehicle when installed or removed. The magnet collects metal particles that can be ground into the vehicle's paint.

Existing Antennas

Care should be taken when mounting a cellular antenna on a vehicle with existing antennas. To prevent any incompatibility problems with two-way or other antennas, the antennas should be mounted no closer than 18 inches. Even with this rule, problems may still occur. Thus, a technican should use a temporary antenna (i.e., magnetic mount) to check for interference before mounting a permanent cellular antenna close to an existing vehicle antenna.

Antenna Gain

Most common cellular antennas are three db gain. This design became the standard because of cellular's low power characteristics, three watts or less. A three db gain antenna will double the effective power output at the antenna. The diagrams below illustrate the effects of increasing the effective power.

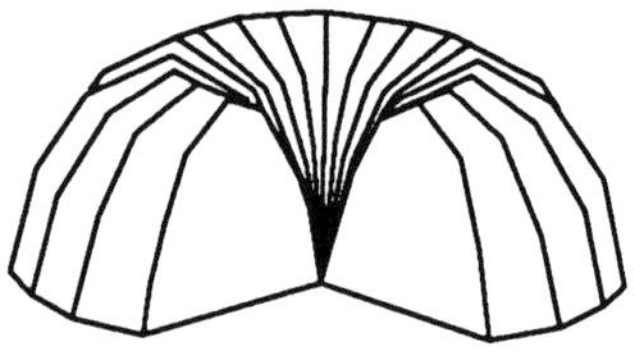

Unity

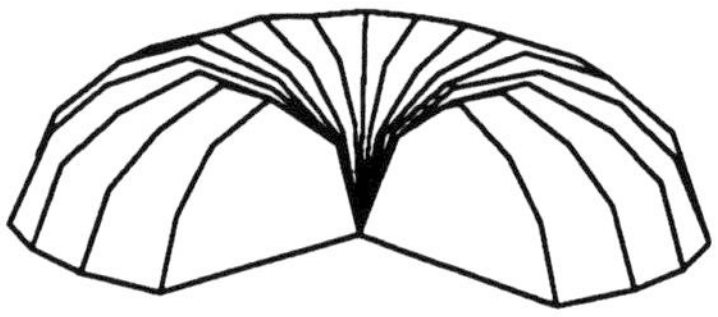

3 db Gain

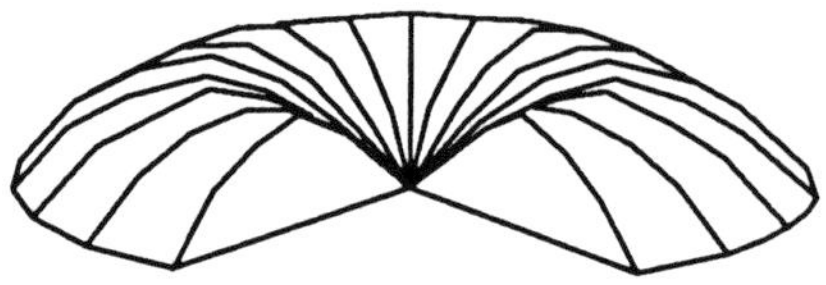

5 db Gain

The antenna's gain compensates for losses caused by the antenna cable, connectors and other antenna components. Each antenna connector induces signal loss. Higher gain antennas cause the signal to become more directional. A high-gain antenna cannot boost the telephone's output power beyond the signal produced by the phone.

ANTENNA LOCATIONS

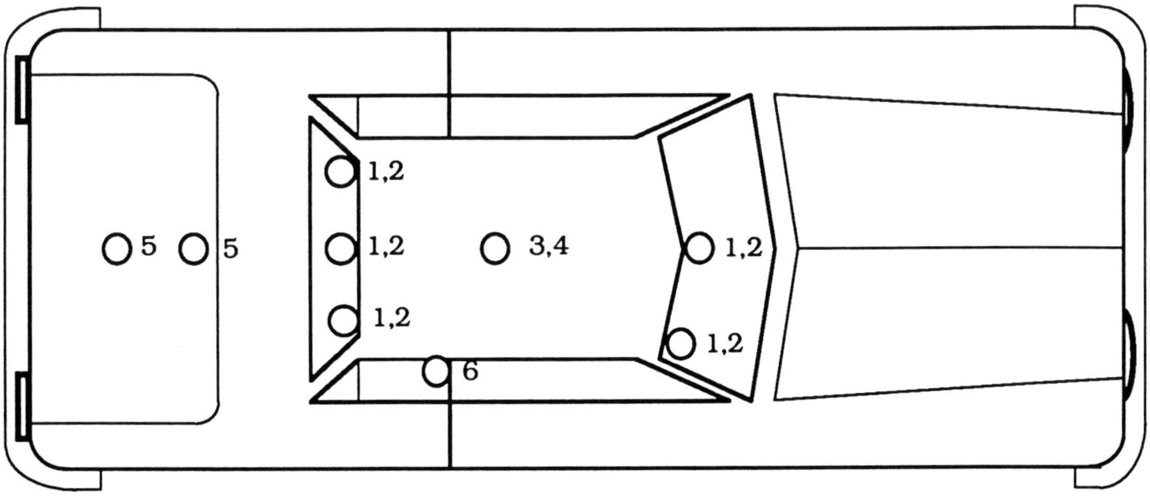

ANTENNA TYPES

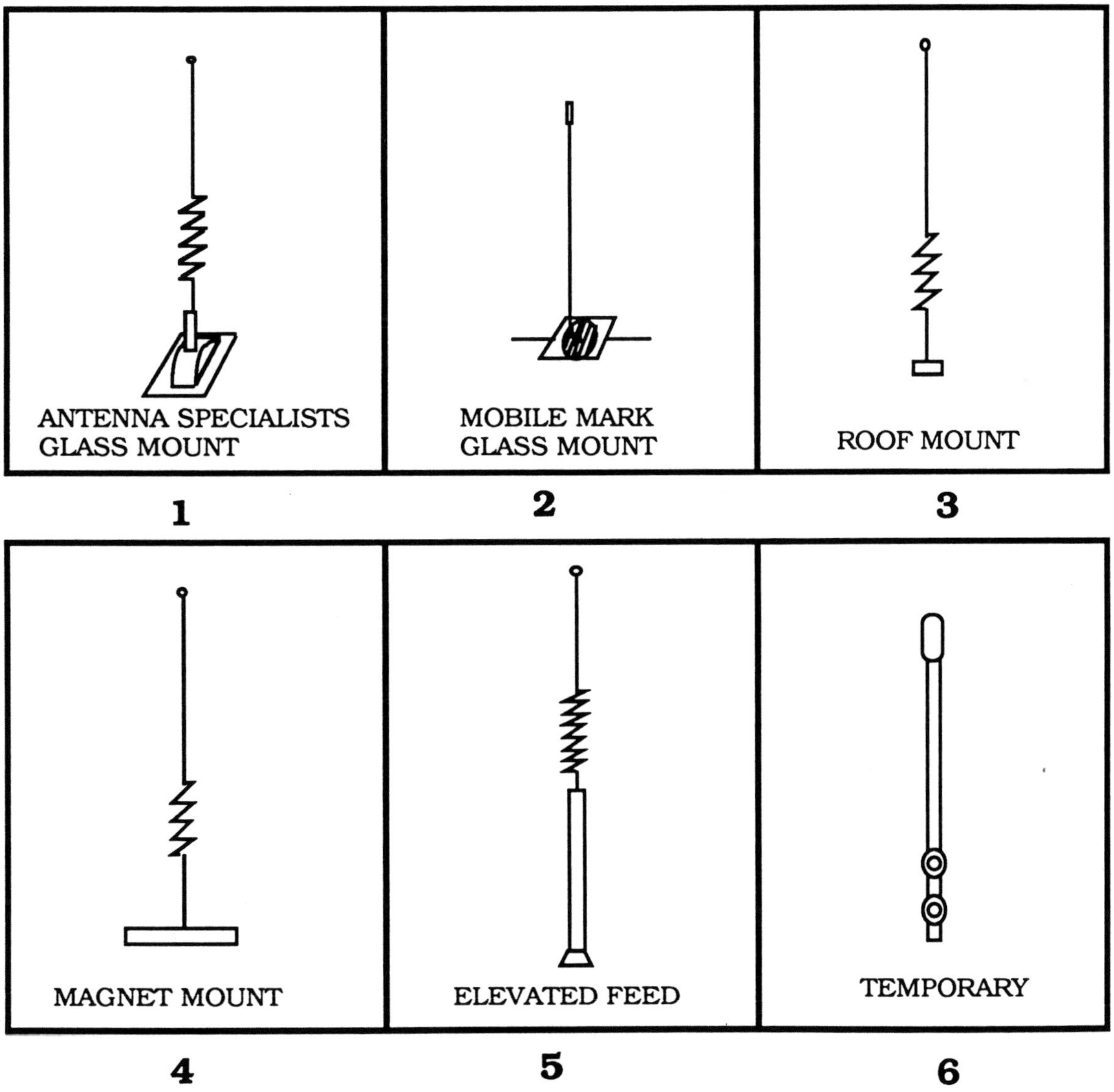

ANTENNA TYPES AND GAIN CHARACTERISTICS

A sampling of typical antennas, applications and gain characteristics is listed below. Since antenna design varies, depending on the manufacturer, it is recommended to review each antenna manufacturer's literature for specific gain information.

Antenna Type	**Gain**	**Application**
Elevated Feed	3 db	Trunks, lumber racks or truck mirror mount. Non-ground plane applications
Glass	3 db or 5 db	Vehicle front, side or rear windows. Designed for standard auto glass.
Marine	3 db or greater	Boats-generally mounted on ratchet mount base
Portable	Unity	Standard equipment antenna for most cellular portable phones
Roof	3 db or 5 db	Vehicle roof
Suction	Unity	Vehicle side window

Antenna Type	Gain	Application
Transportable	Unity	Standard equipment antenna for most cellular transportable phones
Yagi	6 db or greater	Highly directional antenna for stationary applications

5 Telephone Activation

GETTING THE CUSTOMER ON THE AIR

The process of activating a cellular phone is generally transparent to the customer. The customer determines the phone type (i.e., portable, mobile or transportable), the network assigns a telephone number and the technician programs the phone. The process is complete when the carrier activates the phone number. With multiple people and free-thinking computers, it is understandable that "Murphy's Law" occasionally appears.

Choosing Equipment

Telephones come in multiple sizes and colors. All companies, combined, manufacture several hundred models of portables, mobiles and transpacks.

The process of choosing a model can take seconds, minutes, hours, or even months, depending on the customer. I've always figured that customers who take months to choose a cellular telephone must be entertaining when purchasing an automobile.

Depending on the customer's needs, a saleperson can help find the best phone make and model. Portables offer maximum flexibility, mobiles are installed in the vehicle and transpacks offer the advantages of both mobiles and portables. Transpacks are larger than portables and have a higher power output, thus providing a greater range.

Credit Application

If carriers and customers could be perfectly frank with each other, the credit interview would take about one minute.

CUSTOMER: "I'd like a cellular number and I have money."

CARRIER: "That's fine and dandy, but how do I know you'll pay your phone bills?"

The customer would provide a $25,000 letter of credit from Happytown Bank. A handshake would close the deal and they would live happily ever after.

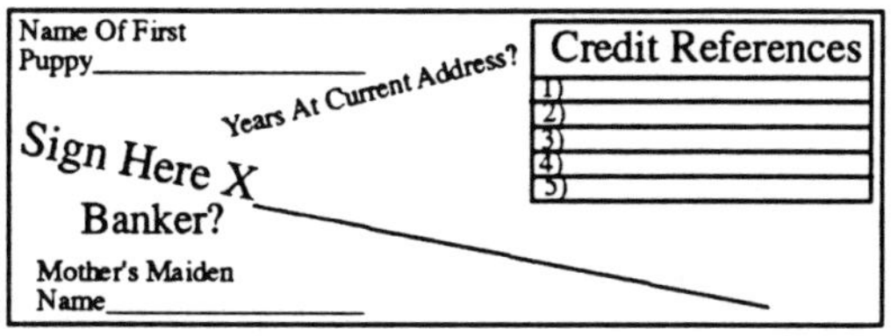

Unfortunately, it's not that simple. Customers complete a credit application to be submitted to the carrier for review. All application blanks must be completed, with the "i's" dotted and the "t's" crossed. Once the customer shows a solid credit record, the carrier welcomes the customer. Carriers also can require an advance security deposit. With the credit application approved, the network gives the customer a telephone number. The carrier tells the technician the number to program into the cellular telephone.

Program Phone

A telephone number must be input into the telephone before the phone will operate. Besides the 10-digit telephone number (area code plus 7-digit number), there are other parameters to program. The system identification number, or SID, is entered to tell the phone which system is "home."

Program Parameters

Mobile Number	Area Code + Phone Number
Unlock Code	3- to 4-digit code
Station Class	Tells system the phones power and channel capibility
Access Overload	0 + Last digit of phone #
MIN Opt	If enabled, the phone sends its 10-digit phone number with each call
System ID	5-digit number that specifies home system

Another item, called Station Class, tells the system about the phone's power output and channel capability. If you have any additional program questions, a complete list of program parameters and the function of each is listed in Appendix B.

Once the technician enters the phone's specific operating information, the shop notifies the carrier of the mobile's electronic serial number (ESN). The ESN is a security number unique to each phone. Networks use the ESN to verify that only one phone is using a specific phone number. A second phone could be programmed to an existing number, yet no calls would connect because of the ESN. Since only one ESN can be activated for each number, the true customer and the pirate could not operate with the same number. If the pirate customer had the system activate the second ESN, the first customer could not use the phone.

Model CEL-28
Serial # 9876554
ESN13300012345

Transceiver Label

Technicians enter telephone program information in a variety of ways. Some telephones program via the keypad. In these instances, the technician enters the program security code and accesses the program parameters. The programming code may be the same for a given manufacturer or it may be a random number installed the last time the phone was programmed.

Keypad Programming

Other models require a special handset only available to certified manufacturer repair facilities. The program handset temporarily replaces the phone's standard handset. With the program handset connected, the program information is loaded into the phone's memory circuitry.

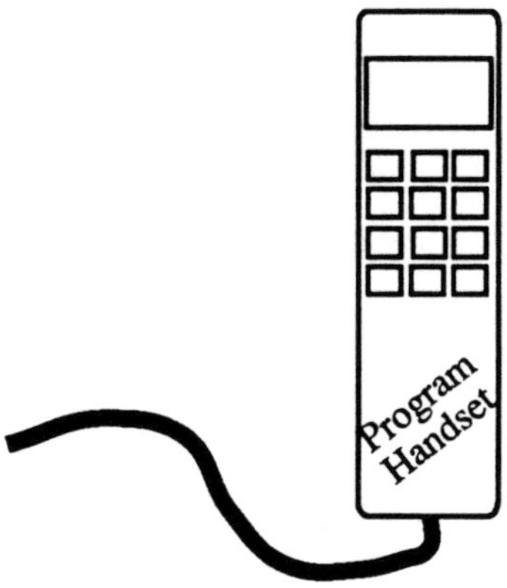

Service Handset Programming

A third type of programming requires a program adaptor. This device plugs into the phone to access program information. With the adaptor in place, the technician enters a sequence from the keypad and the program parameters can be changed.

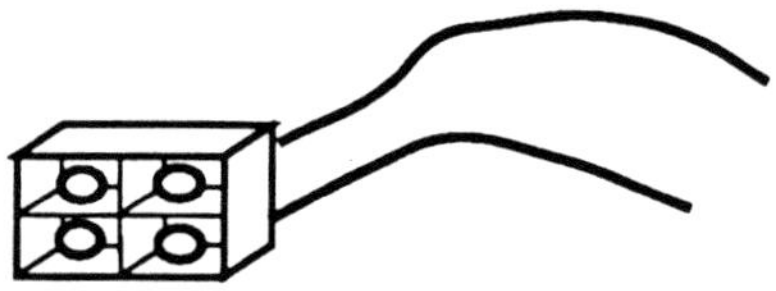

Program Adaptor

A final method of programming is the number assignment module, (NAM). NAMs are small electronic storage devices that are programmed on a machine known as a NAM burner or programmer.

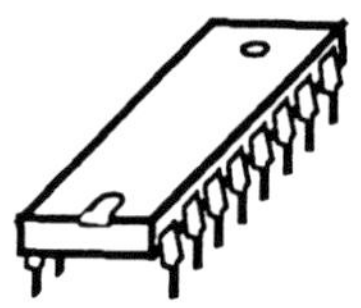

NAM

The phone's parameters are entered into the programmer and a memory chip is inserted into the programmer's output port. The machine's program button is pressed, and voila, the information flows into the chip. This chip inserts into a socket on the phone's logic board and the phone assumes its new identity. The chip method of programming was popular in the mid-1980s.

The largest hurdle in cellular programming is having all programmers, adaptors or service handsets for all models. Customers don't want to hear the "song and dance" about why a shop can't program a phone. This can be a problem for smaller shops, since purchasing all programmers, handsets and adaptors could cost over $10,000.

Check Phone Operation

The number has been assigned, the phone is programmed and the technician may wish to perform a final quality check. This step is called "benching the phone." It's a silly name, yet the process will identify a defective phone before the customer leaves the shop.

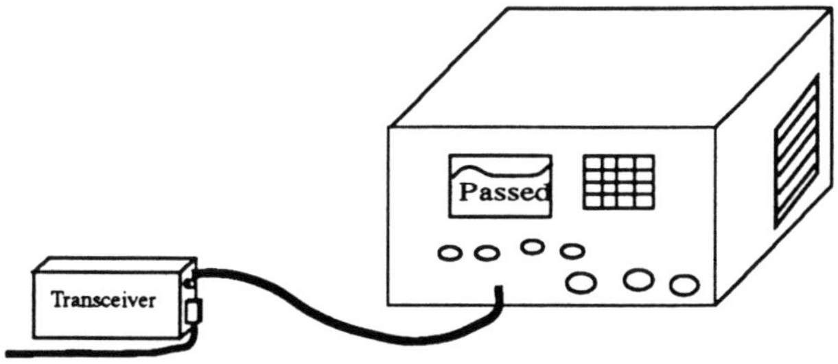

"Bench Testing" With Cellular Monitor

The bench check involves connecting the programmed phone to a cellular monitor. A cellular monitor acts like a closed circuit cellular system. The phone places and receives calls, and simulates handoffs when hooked up to the monitor. During this process, a printer documents the phone's test scores for the technician. If the phone operates within specifications, the phone passes the test. At this point, the technician installs the mobile phone or delivers the portable or transpack to the customer.

Installation (Mobiles Only)

Cellular installation is a multi-step process. The transceiver, handset, antenna, and cabling mount in the vehicle. A detailed explanation of cellular mobile installations is outlined in Chapter 4, *Customer Equipment.*

Phone School

Customers who receive cellular phones without instructions are a liability, not an asset. The new cellular user needs information on telephone features, system options, safety considerations and customer service numbers.

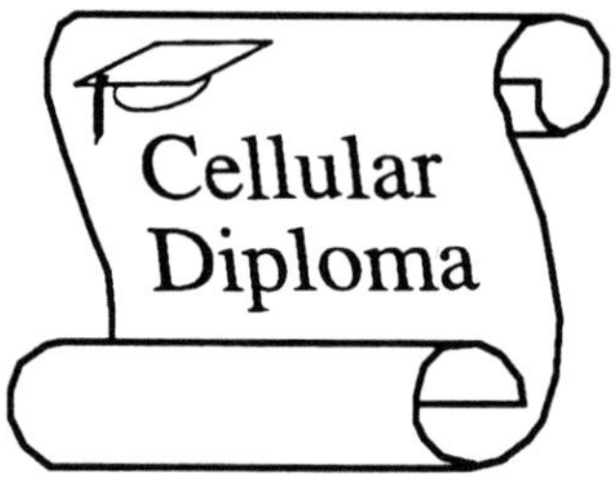

Phone School Graduates Are Better Customers

When teaching phone school, it's easy to detect a customer in information overload. The eyes go glassy and when asked if the information is making sense, the answer is "Huh? Oh yeah, I got it."

A phone school should cover the following topics.

Placing a call
Receiving a call
Using handsfree
Call timers
Locking and unlocking phone
Roaming information
Memory locations
Customer service number
Repair service number
Volume adjustments

Some shops have automated this process on video tape. A customer instruction video can save a technician hours of repeat performances explaining the same phone.

6 Trouble-shooting

TROUBLESHOOTING OVERVIEW

Cellular troubleshooting can be a great lesson in logic. A working knowledge of the system, phone and data provides explanations for some of cellular's wild phenomena. Some sales people describe cellular as "just like your home phone." It sounds excellent in the ads, yet the phone is really a combination of phone, computer and two-way radio. Most symptoms common to any of these three items can occur with a cellular telephone.

One simple way to solve cellular symptoms is to place the problem in one of two areas. Could the symptom be caused by the system, phone or both? If it could be phone-related, what steps can be taken by the user before driving across town to the repair shop? For system-related problems, does the static or dropped call happen in the same area? This may suggest a defective cell or poor coverage area. By asking the right questions, a competent technician or customer service representative can solve many cellular problems without touching the phone.

A reporter asks several questions when covering a story. The purpose is to find the facts. A person attempting to help a cellular customer is no different. Once the facts are known, a solution can be offered. The following questions must be asked by the cellular reporter to figure out if the problem has a simple solution or requires major surgery.

Questions

Who?	Customer's Mobile Number
What?	Phone Make and Model
Where?	Location within the System where Problem Occurred
When?	Date and Time of Problem
How (Often)?	Frequency of Symptom

Possible solutions are outlined over the next several pages for 11 common cellular symptoms. The potential solutions should help solve problems when customers call for service. If these solutions don't solve the problem, it's time for the cellular phone to make a trip to the shop.

The *Troubleshooting Grid* on page 98 simplifies the troubleshooting process. The grid identifies the components associated with specific cellular symptoms.

Fast Busy

Possible Causes

Network: Yes

Phone: Yes

The fast busy occasionally sounds when a call is attempted. The fast busy, also called a directed retry, originates at the network. A customer may receive a fast busy, when dialing, if the nearest cell is handling its maximum capacity of calls. The cellular user dials the phone number and presses the SEND button. The system cannot process the call because of system load, so the system says "try again later." The notification is an obnoxious "beep, beep, beep."

Fast busies also can occur if the registered ESN and the phone's ESN don't match. This varies, depending on the cellular system. Some systems notify the user with a message and others use the fast busy method.

A cell experiencing equipment failure also can cause a fast busy. If the data transmission between the mobile and the cell is not satisfactory, the "fast busy" will sound.

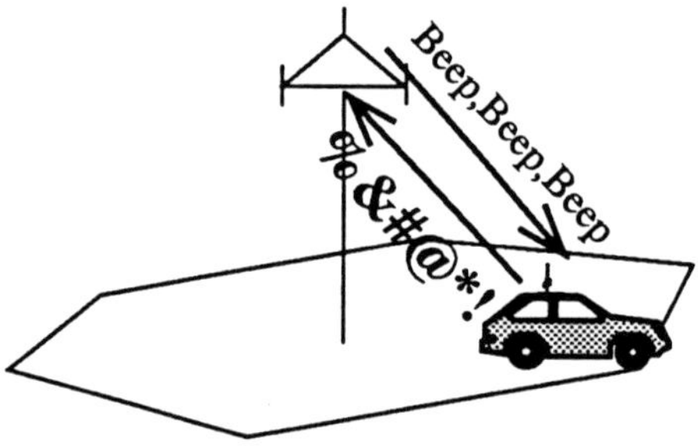

A Garbled Mobile Request Receives a Fast Busy

Another possible cause is a defective mobile phone, where the data scrambles and the system cannot read the mobile's request. This is a typical system response when the mobile is operating in a system "fringe" area. In this instance, the mobile sends the call request to the system. The data is scrambled and sounds like someone talking with their mouth full of mush. Instead of the system saying, "Don't talk with your mouth full," The system tells the phone to generate a fast busy.

Phone Rings-No One There

Possible Causes

Network: Yes

Phone: Yes

Since the phone rings, the ringer volume adjustment is not related to this symptom. The phone's handset volume could easily cause this symptom. By adjusting the handset volume louder, the problem may disappear. It is important to understand that cellular phones generally have three separate volume adjustments: ringer, handset and speaker.

The network may be at fault if the user has activated conditional call forwarding or if the current cell has a defective channel. With the conditional call forwarding, the phone rings 3 to 4 times and, if not answered, the system forwards the call to another phone number. If this feature is activated and the user doesn't answer quickly, the call disappears. It may be interpreted as "Phone rings-no one there."

Many phone components can cause this symptom. The audio must travel as radio frequencies down the antenna and into the transceiver.

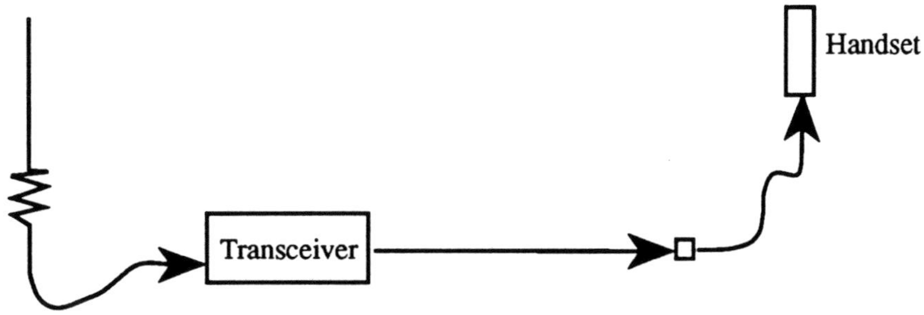

The Cellular Mobile Audio Path

Inside the transceiver, the incoming audio wanders throughout the components and then to the handset via the data cable. The incoming audio can be affected by a defective transceiver, data cable or handset.

Static-Dropped Calls

Possible Causes

Network: Yes

Phone: Yes

"So, John meet me at the . . . SSSSSSSS . . . at 5 o'clock and we'll discuss the. . .SSSSS. . . .Don't forget to bring. . . SSSSSS. . ." (silence). This is a frustrating way to conduct business. If you use a cellular phone, the odds are that some day you will have a dropped call. A dropped call is another word for disconnected call. An occasional dropped call is nothing to worry about. Repetitive dropped calls are unacceptable.

Networks can cause dropped calls if the cellular coverage is weak. Systems with hills, valleys and excessive distances between cells magnify the problem. In a poor coverage area, the radio frequencies have difficulty traveling between the mobile and the cell. If, during a conversation, the system loses the signal returning from the mobile for five seconds, the call disconnects. Besides poor system coverage, a defective channel at the cell site also can cause static and dropped calls.

The point of this explanation is to define the causes and effects of potential cellular problems. Too often, system operators blame the phone. The phone manufacturers blame the networks. The dealers blame both. The cellular users don't care whose fault it is; they just want the problem fixed.

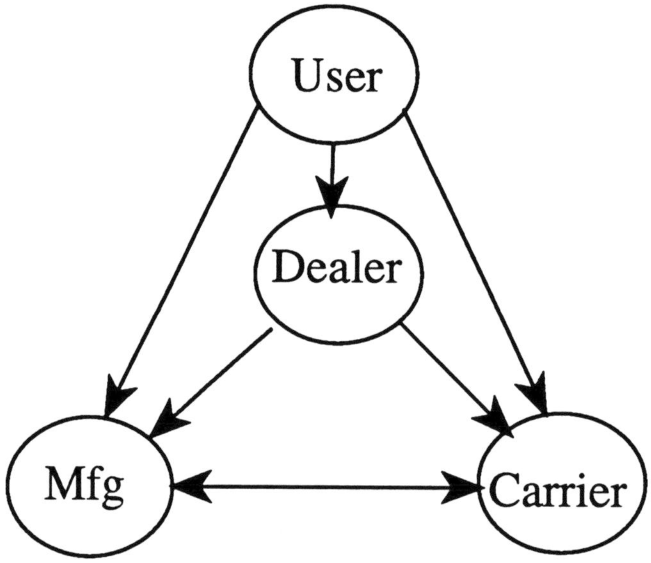

Dropped Call Blame Molecule

Mobiles or portables also can cause dropped calls. Radio frequencies travel from the antenna to antenna cable, to connector and finally enter the transceiver. If any of these components is defective, intermittent reception is common.

The phone's handset and data cable can be responsible for poor reception. Both the handset and data cable carry audio signals to the microphone and speaker in the handset. A broken wire inside the data cable can be the source of poor audio.

Feedback

Possible Causes

Network: No

Phone: Yes

You are stuck in traffic, talking on your cellular telephone ordering a pizza. The main concern is to have the fresh pizza waiting for you when you arrive home in 20 minutes. The conversation is intercepted several times with an obnoxious high-pitched squeal. If the pizza person had "dog like" hearing, the pain would be unbearable.

The high-pitched squeal is called feedback. Feedback happens whenever a microphone can hear a speaker that amplifies the microphone's signal. In other words, it occurs because of a signal loop from the speaker to the microphone. It goes from the speaker to the microphone, from the speaker to the microphone and on and on and on. In cellular, this occurs in the handsfree mode. The handsfree microphone hears the handsfree speaker and sends the signal through the system again. The sound heard by the user is the awful loud howl.

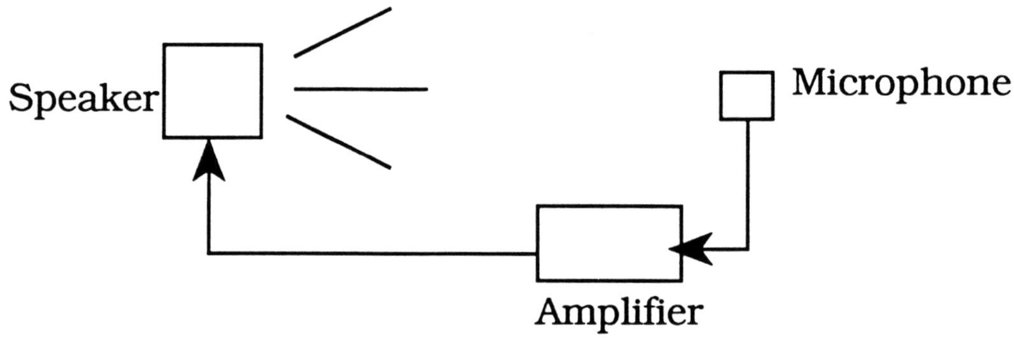

Feedback occurs if the speaker's signal is heard by the microphone then amplified enroute back to the speaker

The problem can be solved a couple of ways. First, try adjusting the phone's handsfree volume. By lowering the handsfree volume, the signal can't be heard by the handsfree microphone, thus no squeal. A second and a more involved method is to relocate the handsfree microphone and speaker. If the two components are mounted too close together, feedback can develop.

Phone Shuts Off Mid-Call

Possible Causes

Network: No

Phone: Yes

A phone that shuts off during a call can be a real communication obstacle. It is important when diagnosing this symptom to be positive that the phone physically shuts off. A good question is, "Do you need to press the POWER button to place another call?" A customer experiencing dropped calls can easily label the problem as "phone shuts off mid call".

Figure out whether you are dealing with a mobile or a portable phone. The best question to ask is, "Is the phone physically mounted in your vehicle?" If the answer is yes, you have a mobile and if the answer is no, the phone is a portable or a transportable. A mobile may experience this problem if the handset coil cord is defective or not securely fastened. A phone with poor vehicle connections will cause this symptom. If a foreign liquid (i.e., coffee or carbonated beverage), spills on the handset/data cable connection, the corrosion will cause many phone problems. Besides "phone shuts off" problems, a foreign liquid can also cause "one way audio and static."

A portable that "shuts off mid-call" can be easily diagnosed as the battery, charger or an internal problem with the portable. If a second battery is available, replace the existing battery. If both batteries operate the same, check the phone's battery charger.

Portable Battery Won't Hold a Charge

Possible Causes

Network: No
Phone: Yes

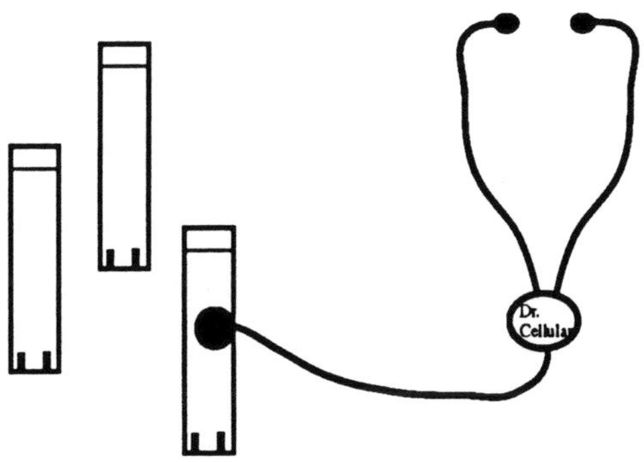

Don't overlook the charger, battery or portable itself as potential causes. As with the "phone shuts off" problem, all three components can be responsible either individually or collectively. If a customer has multiple batteries and each operates the same, the charger or charging method should be suspect. A good rule with nickel cadmium batteries is to discharge the batteries completely before recharging. In this section, we are talking about portable batteries, which are generally nickel cadmium.

If you are dealing with a lead acid (gel cell) battery, the opposite is true. Lead acid batteries should never be fully discharged. Lead acid batteries can be found with some transportable models.

If the batteries and charger have tested as satisfactory, the phone may be the culprit. A portable with an intermittent internal short can drain batteries and shorten talk time. A defective portable should go to a factory-trained technician. The amateur who feels clever enough to repair the portable also should feel rich enough to buy a new one. Internal repairs attempted by users can cause irreversible damage and expensive non-warranty repairs.

No Call Out-European Siren
(No Service Light)

Possible Causes

Network: Yes
Phone: Yes

The user attempts a call and the only sound is one resembling the "high-low" tones of a European siren. The cellular phone produces these tones to notify the user that the phone is unable to place a call. This alert sounds because the phone tries to connect, but the mobile cannot contact the network. The cellular telephone's No Service light usually illuminates when this symptom occurs. You may remember from the network discussion (Chapter 2, *The Network*) that the system sends data to the mobile via a control channel. In the event that the mobile doesn't receive the data signal, the phone's No Service stays lit. This can be caused by the mobile's location, a defective component or a system problem.

A "no service" problem also can be programming-related. When the technician programs the phone, the system identification number, system selection and invalid system ID numbers can be entered. Mis-programming any of these three parameters can cause the No

Service light to stay on. For more information on program parameters, see *Program Terms* (Glossary).

Not Receiving Calls
(Low or No Ring Volume)

Possible Causes

Network: Yes
Phone: Yes

Major frustration! You have distributed your cellular number to several people, they claim to have called, yet the phone never rang. This symptom can be related to volume adjustments, a defective speaker, cabling or a circuit board. Sounds pretty confusing. The most obvious solution is that the wrong number is being distributed by the customer.

Cellular phones, like any other product, can be programmed on any day of the week. A cellular phone programmed Monday morning may be programmed OK, yet when the number is typed for the phone number label, the number is off by one digit. The symptom, no incoming calls, has a simple solution. Recall the phone number from the keypad. Several cellular models will display the programmed phone number simply by pressing RCL + #. Refer to the individual model user's guide for specific operation information. By recalling the programmed number, you can discover quickly if the number matches the labeled number.

If the programmed number is correct, the next step is to check the ringer volume adjustment. Cellular phones have three or four volume adjustments, so make sure you have the correct sequence to adjust the phone's ringer.

Another potential cause for the "no ring" symptom is call forwarding. Call-forwarded landline phones produce a single ring, then the call forwards to the next number. This is the phone's method of saying, "Hey dummy, you forgot to take your phone off call-forwarding."

If the cellular networks offered this service for immediate call-forwarding, the system would be required to make each forwarded call ring once. This would require a voice channel assignment for two seconds and enormous amounts of wasted system time. Thus, with cellular immediate call forwarding, the phone never rings. For "no ring" problems, try to cancel the call-forwarding.

The cellular system also can cause the "no ring" problem. A defective cell or MTSO might not allow the mobile to be notified of incoming calls.

Roam Light Will Not Extinguish

Possible Causes

Network: Yes
Phone: Yes

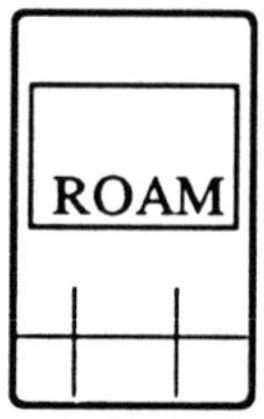

The Roam light is in the cellular telephone display. The light notifies the cellular user when the phone is no longer operating in the "home" system. The home system is the cellular system where the phone was programmed to operate. If the customer is operating in the "home" system and the Roam light is on, there is a problem. On the other hand, a phone programmed for Los Angeles that is operating in Salt Lake City will display the Roam light. For the Los Angeles customer, the phone is operating properly.

The two most important questions when dealing with a Roam light problem are, "What is the 10-digit mobile number?" and "Where is the phone?" The Roam light will glow whenever the system ID (sent by the system), doesn't match the system ID (programmed into the phone).

As mentioned earlier, a customer in the "home" system should not see the Roam light. If the light comes on, the phone is roaming on the competitor's system. The greatest danger of this situation is the potential for conversion. Conversion is a term used to describe a customer being enticed by the competing carrier and lured to disconnect with carrier #1 and join carrier #2. In simple terms, if you were an agent with one carrier and the customer changed services during your commission charge-back period, you would lose money. If everything went according to carrier's plans, customers would stay with the original system forever.

The big question is, how do you keep a phone from roaming with competing carriers in "home" markets? Most phones have a feature, called system select, that solves the problem. System select offers the user multiple system choices. The choices generally include preferred, non-preferred, preferred only, non preferred only and home. An outline of these selections and the effect on the phone is shown below. For best results, it is best to set cellular phones to "preferred only." This reduces any chance of roaming onto the competitor's system.

No Power Up

Possible Causes

Network: No

Phone: Yes

A customer with a “no power up” problem presses the phone's POWER button and nothing happens. No lights, no action, no tones.

It is imperative that you know if the phone is a mobile or a portable. Remember the question, “Is the phone physically mounted in your vehicle?” A mobile phone may have a dead car battery or an unplugged transceiver or handset. For mobile installations you need to know if the other vehicle accessories work. Does the car still start? For portables, did you try another battery? Is the charger plugged into a wall socket that goes off with the lights? Do you understand the charging characteristics of portable batteries?

Cellular Troubleshooting Grid	Activation	Antenna	Battery	Cabling	Call Forwarding	Charger	Coil Cord	Data Cable	Defective Portable	Defective Speaker	Fuse	Handset	Handsfree Circuitry	Microphone	Network	Programming	Ringer Problem	System Select	Transceiver	User Error	Volume Adjust
Fast Busy	*	*							*						*	*			*		
Feedback														*		*					*
No Call Out-European Siren		*							*						*			*	*		
No Power Up			*	*		*	*	*	*		*	*							*	*	
Not Receiving Calls	*				*			*	*	*					*	*	*		*		*
One Way Audio		*		*				*	*	*		*	*	*	*	*			*		*
Phone Rings-No One There				*	*		*	*	*	*		*	*		*				*		*
Phone Shuts Off-Mid Call			*	*		*	*	*	*			*			*				*		
Portable Battery-Dead			*			*			*										*		
Roam Light Stays On																*		*			
Static-Dropped Calls		*		*			*	*	*			*			*				*		

7 Equipment Repair

Why Did It Break?

Even the worst pessimist couldn't predict that a cellular phone was about to break. Many consumer electronic customers expect to buy a product and use it for several trouble-free years. Then reality sets in.

A cellular phone can require servicing just like microwaves, car stereos, televisions and VCRs. Explaining why the phone broke falls in the same category as, "Why do pencil leads occasionally break?" The lead may look fine from the outside and you think you are being careful, then WHAM!, it's broken.

Cellular phones contain many electronic components. They are filled with parts like resistors, integrated circuits, capacitors and whatchamacallits. If any single component quits, the phone can perform some unpredictable tricks.

Setting the Mood

Most cellular repair customers appear surprised that the phone stopped working. The phone has become a crucial artificial limb connected at the mouth and ear. A customer without a cellular telephone equates to a potential tantrum in the waiting room.

A cellular repair facility understands the importance of helping the customer quickly, quoting reasonable repair times and creating an ambience of warmth and compassion.

Customers judge the professionalism of the carrier by the customer service skills at the repair shop. A bad experience at the repair facility may cause a $1,000-per-month customer to jump to the competing carrier.

Most cellular carriers have no problem investing piles of money and professional personnel in the service departments. Technicians with poor customer service skills have no place in this picture.

Experiencing the Problem

The customer describes the symptom and the service writer or technician starts the appropriate paperwork. The service writer asks a series of closed-ended questions. When did the problem first occur? Does the static happen on handsfree and with the handset? Is the problem that they can't hear you or you can't hear them?

The next step is to experience the problem with the customer. This can be difficult if the problem is intermittent. The experiencing step eliminates potential confusion when the phone arrives at the repair bench. Precise communication is essential to solve the problem.

Realistic Time Quotes

A phone headed for service must be treated like a customer's pet. A customer taking a golden retriever to the veterinarian would go ballistic if told, "Leave her with us a few days. Call us Wednesday for a report." Cellular phones are no different. Customers expect reasonable and realistic time quotes. The time quoted can be 20 minutes, two hours or two days. The important element is to offer a realistic time quote before starting the repair. Customer Honesty about time quotes is mandatory. An applicable phrase used by the best cellular shops is, "under-promise and over-deliver." If the technician promises two hours, the repair should be complete with time to spare.

Special Tools

Cellular technicians have a big bag of tricks and several electronic instruments. These include oscilloscopes, voltmeters, wattmeters and service handsets. Each tool has a specific function in diagnosing a sick cellular transceiver. Examples of typical cellular special tools are listed below.

Tool	**Purpose**
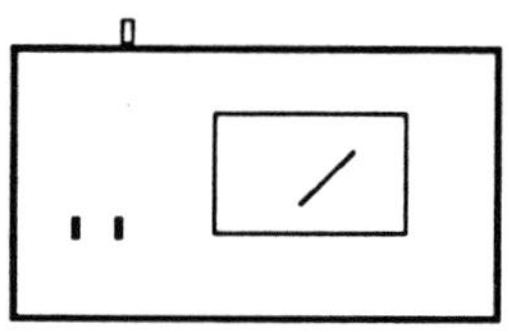 Antenna Tester	The antenna disconnects from the transceiver and connects to this device. The tester sends a small RF signal to the antenna and measures the antenna's performance.

Tool	Purpose
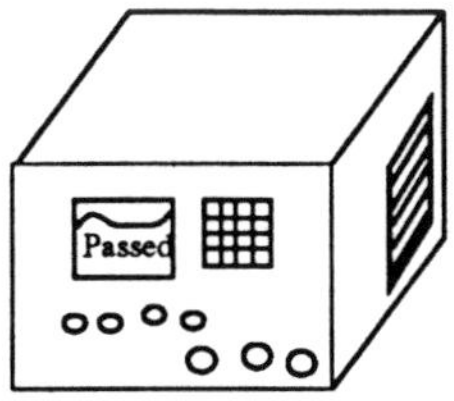 Cellular Monitor	This is the most important tool in the cellular repair facility. The monitor is a "closed circuit" cellular system. The monitor tells mobile phones to receive calls, place calls and experience handoffs. During this exercise, the monitor continually checks the phone for proper operation.
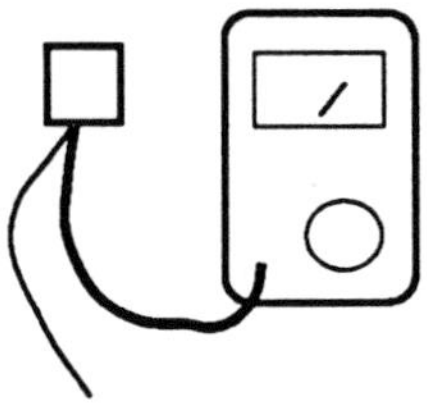 Glass Tester	A glass tester is used when a glass mount antenna is to be mounted on a window with suspected glass impurities. Several vehicle windows have metal particles in the glass for defroster purposes. If a glass mount antenna is installed on this type of glass, the user will receive excessive static and dropped calls.
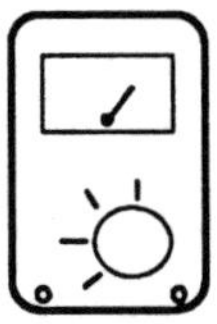 Ohmmeter	An ohmmeter tests the continuity of a wire or other electrical conductor. An antenna cable may be suspected of having a "short" to the vehicle's ground. The ohmmeter can be used to check continuity between the antenna's center conductor and the vehicle chassis.

Tool	Purpose
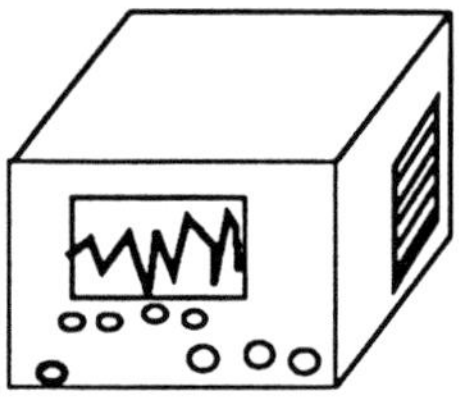 Oscilloscope	The oscilloscope allows technicians to watch things that can't be heard. The scope's display can show the mobile's SAT, voice and signaling tone. The scope's screen looks real high tech.
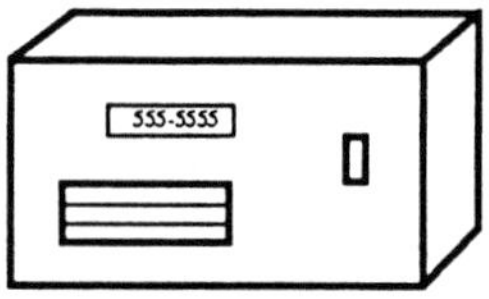 Programmer	Cellular programmers are being used less each year. Most current phones can be programmed from the handset or require a special adaptor. Since there are many older phones still in use, several larger repair shops own a programmer. The machine programs older phones using a small "chip", called a NAM.
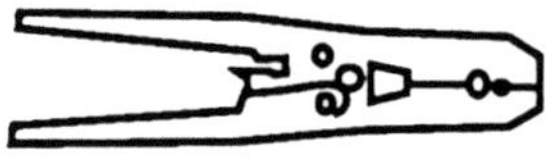 Crimper Tool	The phone's signal leaves the transceiver at the antenna connection. A connector attaches to this output port allowing the signal to reach the cellular antenna. The RF crimper tool installs the RF connector on the antenna cable. RF means radio frequency. If the technician uses the wrong tool, the static and dropped calls will make the customer irritable.

Tool	**Purpose**
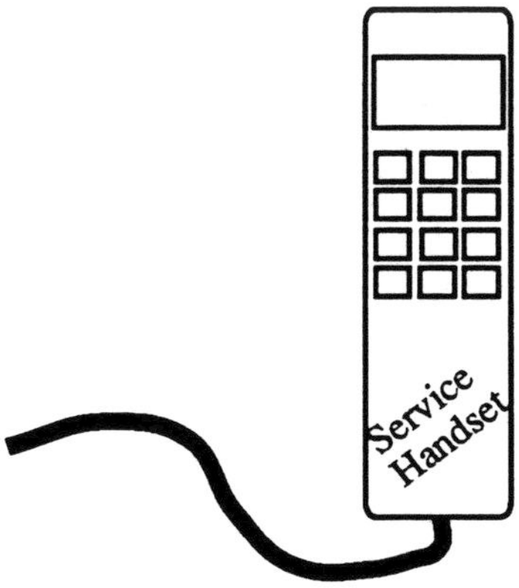 Service Handset	Some phone manufacturers design special handsets for telephone diagnostics. Technicians use these handsets to place the phone into a manual test mode. In this mode, the technican tells the phone to do specific tasks (i.e., power levels or specific channel operation). With this tool, the technician can examine the phone with laboratory precision.
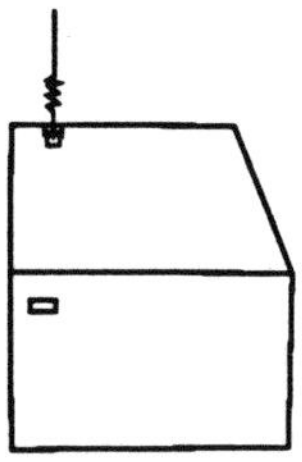 Temporary Antenna	Technicians generally greet the “no service” customer with a temporary antenna. This is the quickest way to eliminate the vehicle's antenna as the problem source.
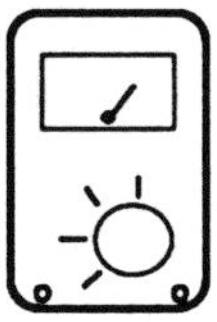 Voltmeter	The voltmeter measures the vehicle and phone voltages at specific test points. The voltmeter is crucial when connecting a power lead to the vehicle. The power source must be strong enough to support the cellular telephone. A voltmeter is vital in deciding where to connect the phone's power lead.

Tool	Purpose
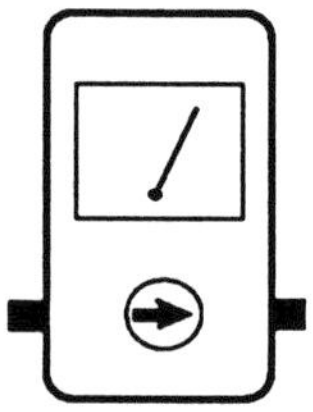 Wattmeter	A cellular wattmeter connects in series with the phone's antenna lead. The cellular signal transmits through the wattmeter on the way to the antenna. As the signal passes through the meter, it measures the phone's power output and antenna operating characteristics. The meter helps to troubleshoot antenna and transceiver problems.

Warranty Versus Non-Warranty Repairs

Cellular telephone repairs fall in one of two categories. Warranty and non-warranty. The paperwork flows according to the phone's warranty status.

A warranty repair is any phone purchased less than X number of months ago. "X" equals the manufacturer's warranty period. If the repair is to be covered by the manufacturer's warranty, the appropriate paperwork must be completed for the specific phone manufacturer. The technician replaces the defective component and the ailing part returns to the manufacturer. The phone manufacturer then repairs or replaces the part, and returns the working module to the repair facility.

The technician generally installs a part from the repair facilities inventory and avoids having the customer wait for the part to return from the manufacturer. The shop bills the manufacturer for the labor and the customer never sees the invoice.

Unfortunately for customers, all repair bills are not paid by the phone manufacturers. An out-of-warranty or abused phone justifies a repair bill.

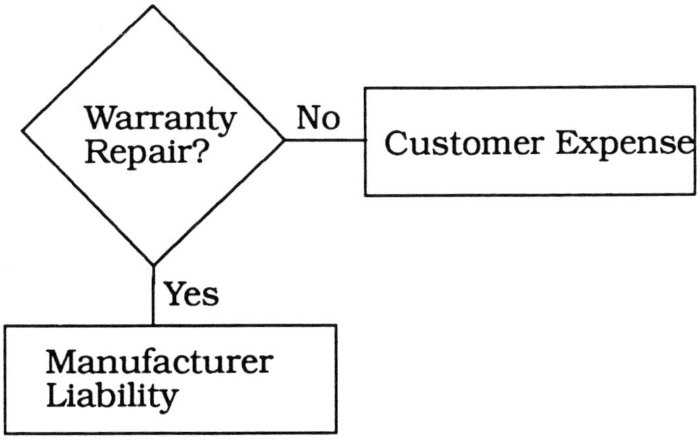

Phone manufacturers won't pay repair bills for a portable dropped into salt water. A portable left on a customer's vehicle roof can take a brutal spill if dropped on the freeway. The freeway drop is also a non-warranty repair. These examples FALL in the category of abuse. For any questions about specific manufacturer warranties, read the warranty card included with the phone.

Non-warranty repairs proceed like the warranty repairs with two exceptions. The first exclusion is that the customer assumes the repair liability. Second, the manufacturer bills the shop manufacturer for repairing any defective boards. These expenses are shared with the customer.

Tracking the Repair

How would you feel if you had a television in for repair at the local TV shop? After two days you were concerned because you missed the "Cheers" reruns. You called the shop for a status report and they couldn't find your television.

When they located your set, the person didn't know if the necessary parts were ordered. Not a pretty scenario! Humans seldom forgive a company that doesn't know the status of the customer's order.

Joe's Repair Service

To combat the potential label of "incompetence," shops have developed tracking systems for a repair's status. The service writer and technician need instant answers to customer questions. Has the phone been looked at? Are parts on order? When should the part arrive? When can I expect my phone? The list goes on and on. The point is, a good tracking system can be a great customer retention tool.

Performing the Operation

Over the years, phone manufacturers have simplified the cellular repair process. The technician removes the defective board and exchanges it for a manufacturer refurbished board. Then the technician verifies the phone operation with the cellular test equipment. Once the technician sees the "one way audio" or the "static" problem has disappeared, the repair is complete.

Notifying the Customer

Calling the customer can be an easy step to forget. Customers love to be informed.

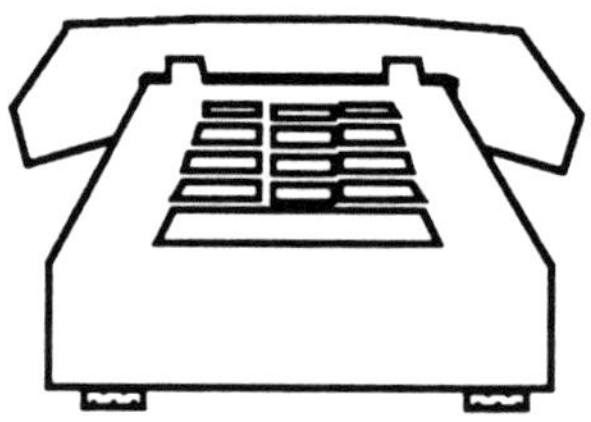

Yet, there is a limit on how much information they want. Most repair customers will ask, "What was wrong with the phone?" An astute technician will simply say, "The logic board was defective, we replaced it and the manufacturer covered the repair." A lost customer, receiving more information than requested, can go into a state known as "technical shut down." In "technical shut down" the customer's speech falters, the eyeballs roll back and there is no chance of returning to an intelligent conversation.

Installing the Repaired Phone

The technician reinstalls the repaired mobile phone and tests the telephone for operation. This is an excellent opportunity to connect the wattmeter to the phone for a final antenna and transceiver check.

As you can see, there are a variety of steps in the repair process. Many steps involve the customer and give the shop an opportunity to make a dynamic impression. The customer's treatment at each step determines if the carrier will keep the cellular user. Can you see why company monies flow freely in service departments?

Part 3

Finding And Retaining Customers

Marketing

Customer Retention

8 Marketing

Cellular telephones enter markets through a variety of distribution channels. Most cellular carriers, generally two in each market, sell equipment (phones) and airtime. The airtime, also known as service, is the billable airtime used by the cellular customer.

Most markets rely on multiple distribution channels to sell the equipment and airtime. A customer may purchase a phone and airtime from the cellular carrier, an agent for the carrier or a reseller. An example of a carrier is CELLULAR ONE or US WEST CELLULAR. A company like Joe's Car Stereo & Cellular would be an agent. An agent is also called a dealer.

Understanding the reseller's position is a bit more complicated. A reseller purchases blocks of numbers from the carrier and handles the customer billing and customer service issues. The reseller idea has never been particularly popular with carriers, possibly because the carriers lack control over the reseller's customer base.

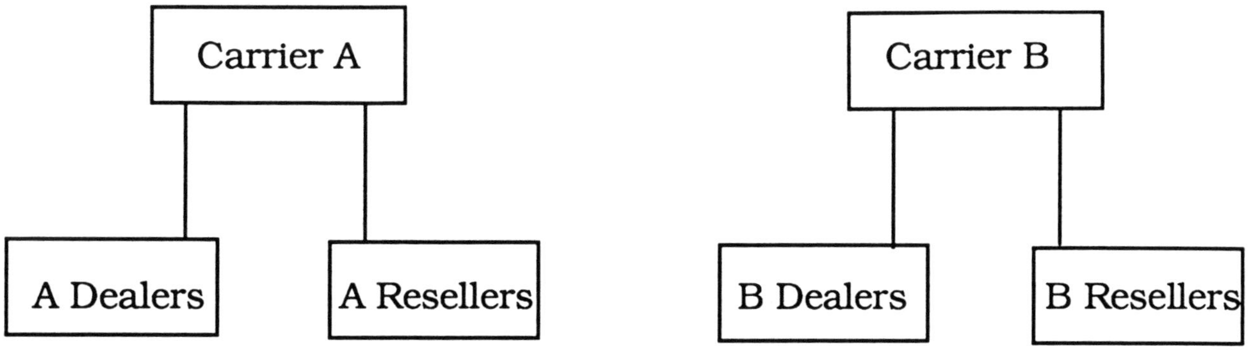

Cellular Purchase Options

A cellular agent or dealer has a different relationship with the cellular carrier than the reseller has. The agent represents the carrier to potential cellular customers. A customer purchases a phone from the agent, completes the carrier's credit application and, if accepted, gets a telephone number with carrier A or carrier B.

Generally, an agent can represent only one carrier in a certain market. Cellular carriers require a customer to prove credit worthiness before establishing a phone number because of the potential airtime and long distance charges. The credit check can amount to minutes or hours depending on the carrier's credit policies.

Once the agent completes the phone activation, the agent's customer becomes the carrier's customer. This doesn't stop the customer from purchasing additional agent products in the future. This is when the customer becomes an asset to the carrier, provided the cellular phone bills are paid.

Cellular Customer Potential

A profitable cellular user potentially could bring a cellular carrier several hundred dollars per month in revenue. The carriers have spent thousands of dollars to answer the following questions. What is the best type of customer? How long does the average customer stay with cellular? Why do customers discontinue cellular service? What are the customer's expectations and perceptions of the cellular service? Is the competition's service better or worse in a particular market? Is the airtime competitively priced? The questions go on and on.

The carrier's main goal is to create a profitable revenue stream for the cellular carrier. Anything causing a blockade in the revenue stream is taken seriously. When a customer begins cellular service, there is usually a "welcome call" from a carrier employee. Customers receive a toll-free customer service number, should any problems arise. If a customer decides to disconnect the phone, a representative will call to find out why.

In a recent demographics study, I saw information gathered by one carrier that helped predict the average attorney's monthly airtime. The study also showed how long the average attorney would continue service with carrier XYZ. These are very powerful statistics if one wishes to predict customer potential.

Bundling

I believe it was the state of California that added the term "bundling" to the cellular glossary. California was the first state to regulate this form of marketing. Bundling occurs when a cellular carrier or agent requires a customer to sign up for airtime when purchasing a phone.

Bundling means: No airtime, No phone, No Deal!

Though this marketing method is common in most states, California wanted this practice stopped. Many dealers fear bundling because the phone sells for less than dealer cost. The dealer subsidizes the sale with the carrier activation fee. Provided the customer stays with the carrier for X months, the dealer retains the commission.

Problems occur when the customer disconnects before the dealer's charge-back period is over. If the customer leaves during the charge-back period, the dealer loses money. It's like doing volunteer work. In this situation, you can do charity work without leaving your current occupation.

How Do Cellular Phones Enter The Market?

Cellular Carriers

Cellular telephone manufacturers use a variety of methods to get the phones to the customers. One method makes the cellular carrier act as the distributor. Cellular carriers have developed a "love-hate" relationship with this distribution system. They love the volume discounts, market control of equipment types and models and the ability to have models "private labeled."

On the down side, there are frustrations related to the equipment business. These include agent credit problems, vendor warranty issues and establishing departments to provide inventory control and stocking problems.

Distributors

Cellular telephones also enter the market through distributors. These companies generally stock limited brands and models. Phone distributors tend to have strict credit policies. I can remember ordering from one distributor who had a simple credit policy. On your first order, it will be sent COD and the rules are cash or cashier's check. The second order had the same rules. I detected they were burned once or twice in the early months.

The distributors purchase the cellular phones and accessories in quantity from the manufacturers. This provides a marginal mark-up as the product is moved to the agent level.

National Activation Service

A clever method of product distribution is a national activation service. These companies usually offer cellular telephones via "mail order" catalogs. The small print states that the phone sells for $XXX.XX and includes activation in your local market. The mail order company establishes contracts with multiple carriers throughout the United States, thus allowing easy activation in the user's home system. This type of activation usually doesn't give the purchaser a choice of cellular carriers.

The service establishes a mobile activation with the A or B carrier, not both, in a particular market. Once the customer purchases the phone, the local cellular carrier pays the mail order company a finder's fee for supplying a new cellular customer.

Auto Manufacturers

Several auto manufacturers have entered the cellular arena. The auto manufacturers have contracts with specific cellular telephone manufacturers. The phone manufacturer produces models to the car manufacturer's specifications. Ingenious cellular telephones have appeared with consumer-catching "bells and whistles." OKI designed a model for CHRYSLER known as the VISOR PHONE. The keypad is located in the driver's sun visor. GM and FORD have cellular models designed by MOTOROLA that offer additional vehicle compatibility and added features.

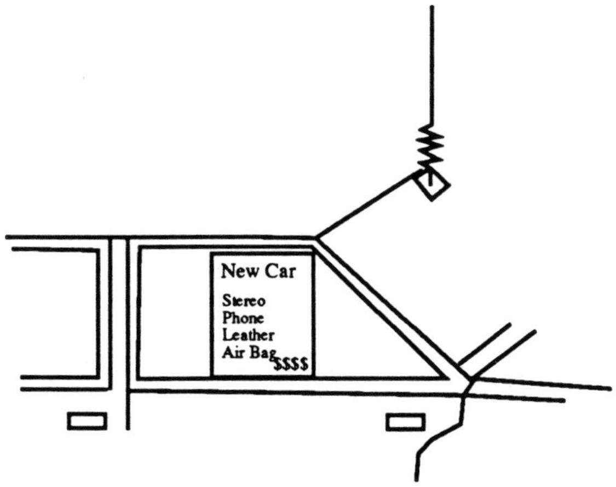

Car dealers offer cellular telephones as an option for most vehicles. Other car models have phones when shipped from the factory. Auto manufacturers have discovered that they can enhance the vehicle's profitability by offering car phones. Besides the phone profit, they can receive a carrier activation fee for providing a new subscriber.

9 Customer Retention

WHY IS A CELLULAR CUSTOMER SO SPECIAL?

Every business relies on repeat customers. Cellular is no different. The cellular telephone industry is dependent on customers because the customer purchases a phone for X dollars, the installation for Y dollars and the big variable is Z, the airtime.

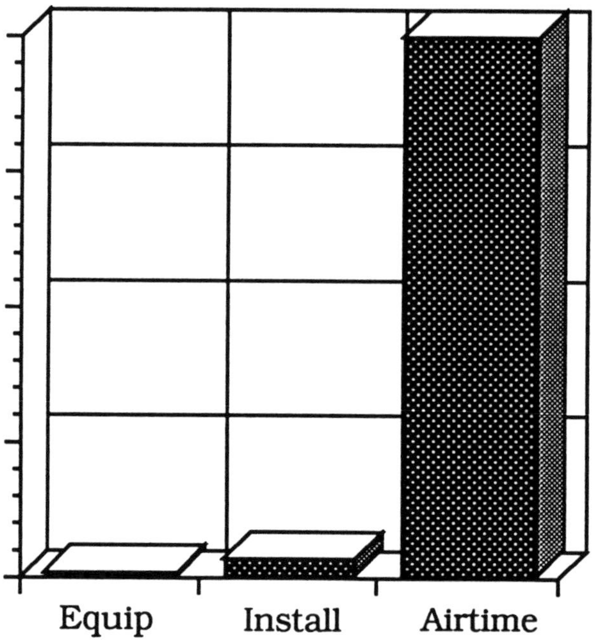

Cellular Profit Margins

The airtime used by a customer can range from pennies to thousands of dollars per month. As with the cable television or paging industry, there has always been customer attrition or disconnects. If a customer disconnects the cellular telephone and was a thousand-dollar-per-month user, the carrier feels the loss. Over the years, the carriers have attributed the disconnects to a variety of reasons. These include sloppy installations, high prices, poor customer service, poor repair service or aggressive customer conversion from the competing carrier.

The customer has the opportunity to meet carrier representatives personnally only three times. The first is when the salesperson signs the customer up for service. The second is at the installation facility, and the third is at the repair shop (if the phone should require service). The remainder of the customer contact occurs via telephone (from customer service) or by the mail (bills). The three situations described above establish the customer's perception of the cellular carrier's professionalism. There must be several carriers who agree with this philosophy as shown by longer agent "charge-back" time and the millions of dollars invested in customer service departments. Several years ago, agents could keep the airtime commission, provided the customer continued service for three months. That figure has grown to six months or more by most carriers. This gives the agent incentive to offer continuing service to the cellular customer. Added service includes installation guarantees and instant solutions to repair problems.

Customer Chatter = Carrier Profits

The successful cellular companies share a simple philosophy. They make most decisions based on a customer's needs. These decisions can lead to such policies as loaner phones, employee professionalism standards, turn-around time, allowing the customer to talk to a technician and repair shop hours.

The information contained in this section should help illuminate that the emphasis needs to be placed in taking care of the customer. There are many excellent books on the subject of customer service.

The successful carriers have followed the advice and role models of customer service pros from a variety of other industries.

Books | Tapes | Video | Classes

Customer Service Training

In dealing with carriers and agents throughout the United States, I have learned about several effective customer service programs. The following pages outline ideas used by successful companies. It is essential when designing a customer service program to remember that the customer starts the revenue stream and can choose to stop the monetary flow at any time. When you consider the cost of finding a new customer, it is clear why so much emphasis is placed on retaining the existing ones.

Customer Service Departments

It's 8:00 PM on a Friday night and Joe Cellular accidently locked his phone. The sales associate told him about the unlock code during "phone school." Unfortunately, Joe forgot that part of the cellular training. Joe is hot, Joe is mad, Joe is hysterical.

He does remember one phone number, the carrier's customer service department. Joe finds a pay phone and a quarter to call the carrier.

From the scenario above, it's obvious that the customer service personnel must be skilled in problem solving, system theory and have a resource library with cellular phone product information. They also need access to customer account information, roaming requirements and knowledge of the problem areas in the network.

Joe's unlock problem may be diagnosed as "Press 1+2+3. Now can you make a call?." In this example it is easy to see the unlimited types of cellular questions the customer service departments answer.

A good customer service representative is similar to a calm and composed air traffic controller. Despite the problem, methodical problem-solving skills and accessible resources can settle the sometimes savage customer.

Identify Front Line Personnel

Besides the customer service personnel, there are many other high-visibility employees: persons checking in vehicles for installation, service writers, sales people, receptionists, and technicians. These employees must have genuine customer respect or it's time to find a new career. The customer satisfaction quotient is determined by their customer service skills, besides performing the tasks of the job.

High Visibility Employees Work In Glass Offices

A customer arriving at a repair shop with a broken phone must be handled with "kid gloves." Both the customer and the telephone are broken. If the repair shop fixes the telephone, yet the technician is rude, wears a ripped shirt and needs a bath, the customer may not come back. Both the phone and the customer arrived broken and the telephone was the only item repaired. The front-line people are the ambassadors of the business and need to be trained accordingly.

Internal Training

Training the key personnel is a constant challenge in the cellular industry. We are dealing with a "high tech" product that needs to be understood and explained in "low tech" terms. The first step in training a front line staff is to hire people who are trainable. Some people cannot work with customers, others consider the upset customer a challenge. A quote attributed to Nordstrom, a Seattle-based clothing retailer, is "It's easier to teach a nice person to do sales than it is to train a sales person to be nice." In other words, the front line must be chosen wisely.

Once the right people are hired, they must be taught the various aspects of the business. The cellular principles are kept in "low tech" terms, since that is how the cellular theory must be explained to the customer. By keeping the principles simple, all employees tell the same story.

I remember a salesperson named Richard. He would ask several departments the same question until he got the answer he wanted to hear. He figured out quickly that cellular's infant stage was similar to Dorothy asking the Scarecrow for directions in The Wizard Of Oz. She asked which way was the Emerald City and the Scarecrow pointed both directions simultaneously. This can be a problem when everyone is reading the same book, but each person sees a different page. Make sure the training programs are simple and consistant.

A list of training sessions for front line personnel is given on the following page. An effective training program for high visibility employees is more than one or two classes and is an investment in the company's future.

Low Tech High Return Classes

Dealing with Difficult Customers

System Theory Cells, MTSO, call processing steps

Tour the MTSO and a Cell

Installation 101 Do a supervised cellular installation

Programming 101 Program a telephone and learn the programming parameters and the effect of each on the phone's performance

Cellular Troubleshooting Take a look at the 10 most common telephone equipment problems. What problems can be solved over the phone?

Role Playing Guaranteed to be a fun exercise. Have employees act as customer service representatives, technicians, customers and other key players.

Phone Skills Is the phone being used as effectively as possible? Are the right questions being asked and the problems solved?

Cellular Repair 101 Outline the cellular repair process. Show the test equipment, board stock and the steps necessary to repair a phone

Bench Check 101 Time for hands-on training with a cellular monitor. By showing the process, it can be explained in simple terms to the customer.

A training program is a continuing process. With a comprehensive training program, each customer will come back, by choice.

Product Knowledge

A novice in the cellular business is easily overwhelmed by the multiple phone manufacturers and the countless models of cellular telephones. Each phone has a specific set of operating instructions and most customers have lost the phone's user guide.

The challenge is to have a working knowledge of general phone features and information on the specific models of phones. A common method to learn phone features is the "Phone of the Week Club".

> Phone of the Week Club
>
> As a member of the Phone of the Week Club, I agree to become knowledgeable about all cellular phones in the Universe.
>
> See back for certification stamps
>
> Member Signature

This customer service group meets weekly to get hands-on experience with new cellular phones. The meeting can include pizza or donuts, depending on the time of the day. Pizza first thing in the morning can make for an interesting meeting.

Earning Trust

You wish to purchase a cellular telephone. The first question from the cellular carrier is, "Can you prove you'll pay the phone bill?." The customer completes the credit application that is processed by the carrier. The carrier's credit thumb either points up or down.

Customers may feel the carrier lacks trust from the start because of the credit process. Keep in mind cellular telephones are often compared to landline telephones in the customer's mind. If someone wanted to switch long distance service to MCI or US SPRINT, chances are the credit process would be quick. The point is that with cellular the customer's integrity is questioned from the beginning

Cellular carriers have thorough credit checks because airtime bills and long distance charges can quickly jump to a considerable pile of money. If the carrier didn't take a few moments for an initial credit check, the losses could be huge.

Once the customer has established a working and paying relationship with the carrier, the concessions can begin. A customer who complains about a dropped call receives account adjustments. A cellular carrier may choose to pay a cellular repair bill for $35.00 to keep a $1,000.00-per-month user. A customer who feels the phone is a lemon may receive a replacement phone. The goal is to keep the customer on the air. A profitable carrier is a happy carrier, and a customer with dependable service is ecstatic.

Field Service

Some carriers offer on-site repair, programming and installations for high volume users. This idea makes life easier for the customer.

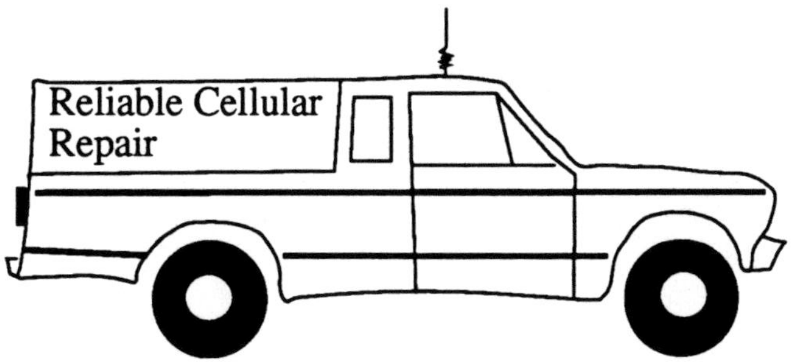

A full-time technician may be assigned to troubleshoot cellular problems in parking garages and residential driveways. Technicians have trucks with a cellular monitor, programmer, replacement PC boards, and an assortment of tools. Often the technician starts the day with no clue of what problems will need attention. Like being a fire fighter, you would never know how many fires to fight until the alarm sounds. This type of program can be expensive, but retaining a high-volume user justifies the cost.

Fast Repairs

Fast is a subjective appraisal. Unfortunately, a cellular customer continues to compare cellular telephones to landline phones. For years telephone companies leased telephones for residential use. If your home phone stopped working, you grabbed the defective telephone and headed to the phone store. You placed the beige beauty on the counter and within a few minutes, the replacement arrived. The exchange phone was taken home, plugged into the wall socket and the repair was complete.

Imagine what the customer thinks if told the broken phone will take 24-48 hours to repair. In comparison to the landline phone, the cellular repair is slow. If a customer were to relate a cellular repair to another consumer electronic product, like a microwave oven, then the cellular repair is quick. Sounds like a variation of the law of relativity. A repair is slow or fast, depending on the reference point.

The speed of cellular repair is a real issue for customers and a continuous challenge for the cellular repair facilities.

Loaner Phones

Several carriers offer loaner phones for customers with telephones confined to the cellular sick bay. The loaner program solves the problem of "Why does the repair take so long?" It allows the customer to continue talking, while the original phone sees Doctor Cellular.

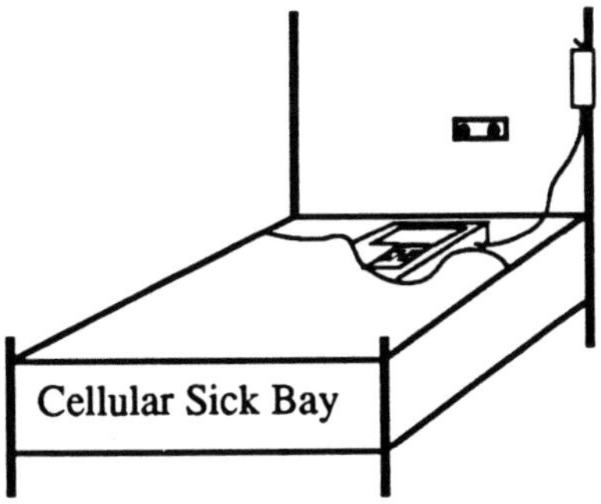

One drawback to loaner phone programs is customers are not always motivated to return the loaner phones promptly. The customer's telephone may be repaired, yet returning the loaner phone takes time. Repaired phones have sat at the shop for several weeks before the customer returns the loaner phone. The good news is the cellular customer continues to use airtime. The bad news is other customers need the loaner phone. My rule of thumb for loaner programs is as follows:

X=Supply of Loaner Phones
1.5 Times X=The Demand For Cellular Loaner Phones

Retail Environment

Retail sales are responsible for an increasing percentage of sales. Many initial telephone sales happened at the country clubs, on sales calls or through referrals. Retail customers are fascinated with cellular's convenience and prestige.

The retail sales of cellular telephones has become a science. Each market has star performer agents who sell most of the phones. A carrier sees a "five phone sales per month" dealer as a liability, not an asset. Instead of establishing 100 "Mom & Pop" cellular dealers, a carrier may allocate 15 solid dealers to attack the market.

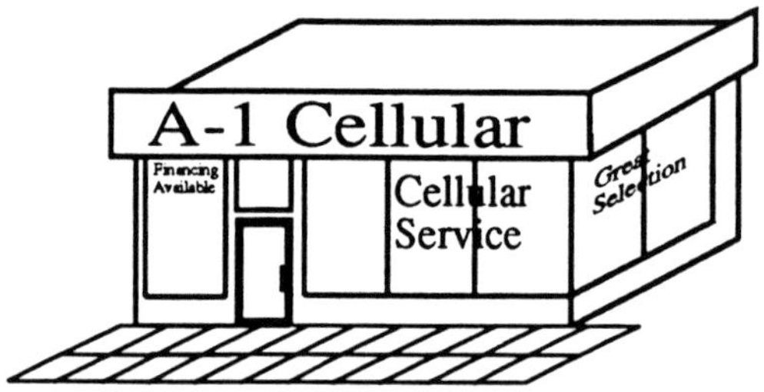

Carriers also have entered the retail business with company-owned stores in high-traffic locations. The lights, fixtures, working displays and knowledgeable salespeople find and keep customers. By establishing a store front, the carrier offers additional customer support. Customers like the convenience and continuity of a storefront. Some agents view the carrier stores as unwelcome competition in the marketplace.

On-Line Support

The best customer service companies heavily rely on data bases. To serve the customer better, customer service departments need instant access to billing information, phone features, phone identification and roaming details.

Valuable time disappears if a representative must call experts for requested information. The information age has trained customers to expect their problem to be solved on the first call, within the first few minutes. We're involved in a fierce marathon in which instant information is crucial for survival.

Consistent Pricing

The market drives price and price drives the market. A new airtime rate may be initiated because of slumping sales. Once the new plan arrives, the market may go crazy. The important element is that when the carrier introduces new pricing, the existing customers must not be forgotten. It may be costly to keep customers informed of their airtime options, yet omitting this step can be expensive.

A customer learning about a missed savings from the competitor is not a happy customer. One theory is that it costs 10 times more to establish a new customer than to keep an existing one.

Recently, I walked into the cleaners and saw a sign,"Shirts 99¢." My shirts had been on the cleaner's rack for 2 weeks. The bill was $8.57 for 6 shirts. When I questioned the clerk about the 99¢ offer, she said the 99¢ special was a new offer and my shirts were at the old price. At this point, I had a choice. Do I wish to continue doing business with an inflexible company?

System Integrity

Cellular carriers spend millions in designing cellular systems. The system geography is mapped and the carrier strategically places cell sites to provide maximum coverage. It always looks better on the drawing board than the finished product does. Once the system is operating, customer complaints highlight any poor coverage areas.

To maintain the best service, carriers are proactive in system improvements. To repair a system "hole", cells can be added or existing antennas redirected. Cellular carriers also use system monitoring devices to measure the system's signal in the entire coverage area. A complete understanding of the system's coverage and necessary improvement is fundamental for customer retention.

Customer Focus Groups

Listening to customers provides a clear insight for a business. Customers eagerly explain how the product should be designed, price considerations and any deficiencies in service.

Carriers Rely on Customer Feedback

It may be a painful but very effective way to learn about the company shortcomings. Customer focus groups provide a customer assessment from someone who has tried the product. Some companies ask customers monthly to assess the organization's level of quality. The results of these surveys often modify policies and produce positive changes.

Customer Follow-Up

Companies with genuine customer concern call customers after a transaction or repair to verify that the product meets the customer's standards. Follow-up calls are essential to notify a customer when special order items arrive or to receive feedback on a vehicle installation.

A follow-up call is an inexpensive way to show the customer the focus is on satisfaction, not on the dollar. My car was in the shop for a simple oil change. Three days later I received a call asking about the service, vehicle cleanliness and timeliness. Three months later I was back for a $300.00 tune up. The phone call told me that the auto dealership valued my business and wanted me to return in the future. A company's greatest asset is a group of customers who continuously talk about the wonderful service.

Part 4

Reference Materials

Appendix A
Phone Features

CELLULAR TELEPHONE FEATURES

Depending on the make and model, cellular telephones have a multitude of features. The following list is a helpful overview of phone features found in the cellular industry. Since the different telephone manufacturers have unique names for the phone features, the list should provide a simplified look at these features.

A/B SELECTABLE CONTROL: See System Select.

ACCUMULATED CALL TIMER: See Resettable or Cumulative Call Timer.

ALERT VOLUME: See Ring Volume.

ALPHANUMERIC ERASE, RECALL, SCROLL AND STORE: This feature allows user to store a name and a number in a specific memory location. Especially helpful for phones having over 10 memory locations because of the user difficulty in remembering which number is in a particular location. As the sub-headings imply, "erase" clears the memory location, STORE and RECALL are used to load and review memory at a later time and the "scroll" feature allows the user to step through the memory locations.

ALPHANUMERIC SCAN: See Alphanumeric Erase, Recall, Scroll and Store.

AREA CODE PROGRAMMING: The mobile phone number is a ten-digit number including the area code. With this feature, the user can re-program the first three digits of the phone number from the handset.

AUDIBLE CALL TIMER: This timer signals the user when the conversation has passed another minute or other preset increment of time. The audible call timers are generally heard only by the cellular user.

AUTOMATIC LOCK: When activated, this feature causes the phone to lock electronically whenever the cellular phone is powered on.

AUTOMATIC POWER OFF: Designed to prevent the phone from draining the car or portable battery. This feature will automatically power phone down after a pre-selected duration if the phone is not used or placed in the horn alert mode.

AUTONOMOUS REGISTRATION: If activated, the phone will register its location to the cellular system. Once the telephone has registered, the system would know that the mobile operating within proximity.

AUTO REDIAL: When activated, the phone will automatically redial a number that was previously denied because the system was busy or the mobile was operating in a fringe area and received a reorder from the carrier. Some models will also redial on calls receiving a busy signal.

AUTO RETRY: See Auto Redial.

AUTO STORE: See Memory Auto Load.

AUXILIARY ALERT: See Horn Alert.

BACK LIGHTING-ADJUST: Allows user to set back lighting in the handset display. A useful feature for liquid crystal displays.

BEEPER TIMER: See Audible Call Timer.

CALL DELAY: When activated, the user enters a telephone number that can be dialed later, usually within minutes.

CALL DURATION ALERT: See Audible Call Timer.

CALL DURATION DISPLAY: See Individual Call Timer.

CALL IN ABSENCE INDICATOR: This feature, when enabled by the customer, or automatic with some models, notifies customer of any unanswered incoming calls.

CALLING CARD MEMORY: See Credit Card Memory.

CALL MAIL BOX: See Call In Absence Indicator.

CALL RECORDING: With this feature the user can record conversations, eliminating the need to take notes while in conversation.

CALL RESTRICTIONS: Allows user to program various incoming, outgoing and memory location restrictions to prevent unauthorized persons from using the telephone's full call potential.

CALL SCREENING: If activated, the user will receive a displayed indicator if an incoming call is attempted. The phone will not ring and the calling party will not know that the mobile has been paged.

CHANGE-IN-STATUS INDICATOR: Notifies the cellular user when the phone has gained or lost service or the phone has entered an alternate cellular system.

CRADLE AMPLIFIER: This amplifier can be switched on and off by the user to save the phone's battery life.

CREDIT CARD MEMORY: These specific memory locations are used for credit card numbers. Generally an alternate long distance carrier will have an access number (1-800) and the user will be given a personal security code. To place a call the access number must be first completed, then a tone is heard. At this point the user must dial the telephone number called and the personal security code. Also see Memory Pause, Recall and Store.

CUMULATIVE CALL TIMER: This timer is similar to an odometer in that the counter cannot be reset. Some users record the timer amounts during certain time periods to provide an approximation of the airtime used.

DELAYED CALLS: See Call Delay.

DIRECTORY SCAN: See Memory Scroll.

DISCONTINUOUS TRANSMISSION: A power-saving mode of operation for portables developed by MOTOROLA. Also called VOX operation. To operate in this mode, the phone must be equipped (and enabled by user or dealer) and the cellular system must be equipped.

DISPLAY CHANNEL: When activated, cellular telephone displays current channel phone is using.

DISPLAYED CALL INDICATOR: See Call In Absence Indicator.

DISPLAY MODE: Allows user to select the maximum number of digits that can appear in the display.

DISPLAY NUMBER OF CALLS: See Total Calls Placed and Received.

DISPLAY OWN NUMBER: Allows user to view cellular number programmed into telephone. A handy feature if the phone has multiple phone numbers thus the current number can be verified.

DISPLAY SYSTEM ID: A method for the customer to verify the system ID programmed into the phone.

Display Your Telephone Number: See Display Own Number.

DTMF Duration: Some answering machines or message retrieval systems require a longer keypad tone than the standard 95 milliseconds. With this feature the user may program, from the keypad, the phone's tone duration.

Earpiece Volume: See Handset Volume.

Electronic Locking: See Lock or Partial Lock.

Electronic Lock Mode: Allows user to select lock mode call restrictions.

Error Tone: Provides user with an error tone if the keys have been pressed in the wrong order.

ESN Read: Permits user to view electronic serial number of telephone. This information may be requested if the cellular phone is attempting access to an alternate cellular system.

Function Menu: See Option Review.

Handset To Handsfree Operation: Some phones require an extra step before replacing the handset on the cradle. On these phones, if a customer wished to transfer a call from the handset to handsfree mode and the handset were simply placed on the cradle, the call would end.

Handset Volume: Adjusts the earpiece volume to control the receive audio level for calls in progress.

Handsfree Call Initiation: Most phones initiate handsfree calls simply by leaving the handset in the cradle and placing a call. Some models require an extra step (i.e., press SEND button twice) to initiate this process.

Handsfree Volume: Adjusts the receive audio volume for handsfree operation. Also see Speaker Volume.

Help Mode: When activated, the phone will automatically send the telephone number four seconds after the seventh digit has been entered.

Hidden Credit Card Location: See Secret Memory.

HORN ALERT: Provided the phone's horn alert circuit has been connected to the vehicle, this feature will activate the horn or lights when calls come in. It must be activated by the user before turning off the ignition switch. The horn alert will not sound unless the ignition has been turned off.

HORN ALERT POWER SAVE: See Automatic Power Off.

HOT KEY DIAL/LOAD: This feature allows user to store a frequently dialed number in the "Hot" location. The number may be recalled and sent simply by pressing a single key.

INCOMING CALL INHIBIT: When enabled, no incoming calls are allowed.

INDEPENDENT/DEPENDENT CRADLE: This user programmable feature determines if the call will be ended by replacing the handset on the cradle. The phone should be programmed "Independent" if the phone has handsfree.

INDIVIDUAL CALL TIMER: Measures and displays the duration of the call in progress or the previous call.

KEY BEEP MUTE: See Silence Scratch Pad.

KEYPAD VOLUME: This adjusts the tone heard whenever a key is depressed on the phone's keypad.

LAST NUMBER REDIAL: Permits user to recall previous number called and dial it again.

LCD ANGLE: Used for adjusting the angle of the liquid crystal display for maximum user visibility.

LOCAL USE: This option is generally programmed in the NAM. Some phones make this option user programmable. When enabled, the phone is permitted to use the local options available for that particular cellular system

LOCK: Allows user to electronically lock the phone to prevent unauthorized use.

LOW BATTERY INDICATOR: Alerts user when battery needs recharging. This is an important feature because some batteries can be severely damaged if deep discharged.

MANUAL REGISTRATION: See Autonomous Registration.

MASTER RESET: With this feature the phone's options and memory locations may be reset by a simple keypad sequence instead of clearing each option.

MEMORY AUTO LOAD: Used for storing numbers in the telephone memory locations. With this feature the number can be entered and stored in the next available location.

MEMORY AVAILABILITY: Displays to customer how much of the phone's memory is still available.

MEMORY NAME TAG: See Alphanumeric Erase, Recall, Scroll and Store.

MEMORY PAUSE RECALL AND STORE: Vital if customer uses a telephone credit card. Numbers can be stored to provide automatic pauses when they are recalled. Thus an access number, personal identification and telephone number may be stored in the same memory location. To dial the number, the customer would recall the memory location and release each portion of the number stream at the appropriate time.

MEMORY PROTECTION: Prevents memory locations from accidentally being erased.

MEMORY SCAN: See Memory Scroll.

MEMORY SCROLL: Allows user to step through the memory locations with a single key depression. Used for reviewing memory contents.

MESSAGE RECORDER: Similar to an answering machine. Incoming missed callers may leave a message for the cellular phone user. To use this feature the phone is generally left on or in the horn alert mode.

MICROPHONE AUDIO MUTE: See Mute/Hold.

MUTE/HOLD: Places call on hold to permit cellular user to converse with others in the vehicle without the risk of the called party hearing the conversation.

MULTIPLE NAM SELECT: Provided the cellular phone is equipped and programmed with multiple numbers, this feature allows user to select which telephone number to use.

NEW UNLOCK CODE: A keypad programmable feature allowing the user to re-program the current unlock code, eliminating the need to have the phone reprogrammed by the dealer.

NUMBER SCROLLING: See Memory Scroll, Recall and Store.

OPTION REVIEW: A handy feature if the phone has multiple options. The user may simply enter this mode and scroll the options to show their status.

PACKED MEMORY: See Memory Pause.

PARTIAL LOCK: Similar to normal lock or full lock. Partial lock may still allow incoming calls or certain outgoing calls.

PERMANENT MEMORY: This number is generally stored in the NAM when the phone is programmed. This location is available to the user for recall only.

PLACING CALLS USING MEMORY: See Recall From Memory.

POWER ON/OFF: Turns phone on and off.

PROGRAMMABLE AREA CODE: See Area Code Programming.

RECALL FROM MEMORY: Once the frequently used numbers have been stored in memory locations, they can be recalled and sent with fewer keystrokes than if the entire number were dialed.

REPEAT DIAL: See Auto Redial.

RESETTABLE CALL TIMER READ, REST AND SET: This timer permits user to review the approximate air time used. Some models have a call timer increment set mode to present the timer increment (i.e., 30 or 60 seconds) to match the billing increment.

RING VOLUME: Adjusts the volume of the audible alert tone to notify cellular user of an incoming call. Also called Alert Volume.

RINGER SILENCE: Programmable feature allowing user to deactivate the ringer.

RINGER TONES: User may select different alert tones.

ROAM INHIBIT: When enabled, the cellular phone will not go into service in any systems other than the programmed home system.

SCRATCH PAD MEMORY: Allows user to enter and store numbers while a call is in progress.

SCROLL DISPLAY BACK/FORWARD: Often the telephone number dialed is beyond the capacity of the handset display. With this feature, the display can be scrolled to view the complete dialed telephone number.

SECRETARIAL MEMORY: See Temporary Memory.

SECRET MEMORY: These memory locations may be viewed and used only with a special access code.

SELECTABLE CALL TERMINATION: See Independent/Dependent Cradle.

SELECTABLE SYSTEM REGISTRATION: See System Select.

SIGNAL STRENGTH INDICATOR: Provides user with a visual indicator of the cellular system's signal at a particular location.

SILENCE KEYPAD: Allows user to disable the keypad tones heard whenever a number key is depressed. This feature is active whether or not a call is in progress.

SILENCE SCRATCHPAD: With this feature the user can disable the feedback tones heard during a conversation when the number keys are pressed.

SILENT CALL: See Ringer Silence.

SPEAKER VOLUME: This adjustment varies the audio heard from the cradle or optional speaker. The ring volume is generally a separate adjustment.

SPEED DIAL: A faster way to recall a number from memory than the conventional method. When a number is sent using speed dial, the number is generally not displayed before placing the call.

STOP WATCH: Allows the telephone to act as a stop watch.

STORING IN MEMORY: Numbers can be stored in the phone's memory locations for easy future retrieval. Also see Recall From Memory.

SUPER SPEED DIAL: See Speed Dial.

SYSTEM SELECT: The cellular markets are divided into two systems, non-wireline (A) and wireline (B). With this feature the phone can scan wireline then non-wireline or vice versa. Other options include home only or preferred system only. Special attention must be given to this option if the customer roams frequently on alternate systems while traveling.

TEMPORARY LOCK CODE: This code can be programmed into the cellular telephone if it is to be loaned and the owner wishes the standard unlock code to remain unknown.

TEMPORARY MEMORY: Telephone numbers can be stored in these locations for short periods of time without cluttering the normal storage addresses.

THEFT ALARM: Before leaving the vehicle the user must activate this option. If an unauthorized person turns on the telephone without entering the unlock code, the cellular user is notified at a pre-programmed telephone number.

TIME BEEP: See Audible Call Timer.

TIME OF DAY SET: Allows user to set the clock in the telephone display.

TONE DIAL FROM MEMORY: A handy feature when multiple sets of telephone numbers or access numbers are required (i.e., calling cards). Once the call is placed, a second number may be recalled from another memory location and dialed during the original call.

TOTAL CALLS PLACED AND RECEIVED: This counter displays the quantity of calls outgoing and incoming for the cellular phone.

UNANSWERED CALL INDICATOR: See Call In Absence Indicator.

UNLOCK: Sequence required to unlock phone once the phone has been electronically locked.

VIEW DISPLAY: See Scroll Display Back/Forward.

VOICE ACTIVATION: An added feature to the memory locations. The user recalls telephone numbers simply by speaking a pre-programmed name into the microphone. The number is then recalled from memory and automatically dialed.

VOICE ANNOUNCEMENT: See Message Recorder.

VOX: See Discontinuous Transmission.

Appendix B

Program Terms

During the program and activation phase, cellular telephones are programmed with the information listed below. This glossary was assembled to explain the program parameters and the effect each has on the phone's performance.

	Standard
ACCESS OVERLOAD CLASS: This is a number (0-15) that specifies the level of priority the mobile has when accessing the system. This parameter was designed to give all users equal opportunity to access the system in high usage situations. Consult the customer's local cellular carrier for this number. There is no relationship between numbers and priority (i.e., higher numbers do not necessarily mean higher priority).	**XX**
AIR TIME COUNTER: Activates call timer and gives customer ability to monitor airtime usage. This feature usually rounds up to the next minute and adds time even when called party doesn't answer or line is busy. The feature should be used as a rough estimate of airtime used.	**Enable**
AUTO RECALL: Recalls last number dialed. A handy feature if you call a busy number and wish to redial. The number can be dialed a second time simply by pressing the SEND key on the handset.	**Enable**
DIVERSITY: A feature available on certain phones with two antenna ports and receiver sections. One antenna is for receive and transmit, with the other connector for receive only. If a mobile phone has the diversity option, two antennas are mounted on the vehicle. Very few phones have this feature.	**Disable**

	Standard
END-TO-END SIGNALLING: When enabled, the mobile can transmit DTMF tones on the voice channel. DTMF tones sound each time a number key on the handset is pressed. The keys, if pressed during a call, can be used to access long distance services, answering machines and other DTMF activation devices.	**Enable**
GROUP ID: This is a number (0-15) that specifies how many significant bits of the System ID are compared when processing local control orders. Consult the local cellular carrier for this number.	**XX**
HANDSFREE AUTO MUTE: When enabled, handsfree is always muted on incoming calls. The user can un-mute the microphone from the handset to complete the call. This feature may be handy if operating the cellular phone in a high noise environment.	**Disable**
HANDSFREE OPERATION: This option enables the handsfree portion of the equipment. Some phones with handsfree are still programmed with the handsfree option disabled. For best results, refer to the phone manufacturer's specifications.	**Disable**
HORN ALERT: When the horn alert option is enabled, the user may have an external device (i.e. vehicle horn, headlights, etc.) signal an incoming call while away from the vehicle. Without the horn or lights connected, this mode also allows the customer to leave the mobile on to receive incoming calls without having the ignition "on."	**Enable**
INITIAL PAGING CHANNEL: A three digit channel number specifying the system control channel to start the scan. The initial channel for system A is 333. The initial paging channel for system B is 334. Each carrier (A or B) has 21 paging channels (also called control channels). The channels for the A carrier start at 333 and end at 313. The paging channels for B carriers are 334 to 354.	**333 or 334**

	Standard
INVALID SYSTEM ID: System ID of competing carriers can be programmed into the phone. The numbers are used by the phone to block all roaming on specific systems. This feature is found on limited models of phones. It is an important parameter to change if the customer is switching carriers within a market. Failure to change the invalid system ID in this example would cause the phone to display "no Service."	**XXXXX**
IMTS COMBINATION: Enable this feature when the phone has dual system capability (IMTS and cellular). To program the IMTS portion of MOTOROLA phones, a special MOTOROLA adaptor is necessary. IMTS, also called trunking, is not compatible with cellular.	**Disable**
LOCAL USE: When local use is disabled, the mobile will not respond to any local control orders received from the system.	**Enable**
LOCK MODE: Allows phone to be locked by customer. The lock feature is an electronic lock.	**Enable**
MIN OPTION: When the MIN mark is enabled, the mobile will automaticaly send its area code to the system when originating or answering a call. When disabled, the mobile will send the area code when the cell site requests it. MIN is an abreviation for Mobile Identification Number. See Phone Number.	**Enable**
PHONE NUMBER: A ten-digit number identifying the mobile phone (three-digit area code and seven-digit telephone number). Also called mobile identification number (MIN).	**XXX-XXX-XXXX**
PREFERRED SYSTEM: This refers to the mobile's home system type. This parameter determines which system to scan first. The preferred system is the customer's "home" type system. System A denotes the non-wireline systems and System B specifies the wireline networks.	**A or B**
REPERTORY OPTION: When this option is enabled, phone numbers may be stored and recalled for simplified dialing. Also called the abbreviated memory.	**Enable**

	Standard
SCRAMBLER OPTION: This feature requires additional hardware in phone and at called party's phone. The signal is encoded and decoded by each party, making conversation monitoring difficult.	**Disable**
SECURITY CODE: Generally a four-to six-digit number programmed into NAM allowing user access to restriction levels, unlock code and other features.	**XXXXXX**
SELECTABLE SYSTEM PROGRAM: Allows user to select wireline or non-wireline carrier from the handset. It is a useful feature when roaming in non-home systems. With the system select feature the customer can generally select between A Only, B Only, AB, BA or Home.	**Enable**
SERVICE LEVELS: Restriction levels programmable on certain phones to limit types of calls (i.e., outbound calls only from memory location 01-03, no long distance or area code restriction). Service level four on MOTOROLA phones denotes no restrictions.	**4**
STATION CLASS: This is a number (0-15) that specifies the power class and type of cellular telephone equipment. For mobile equipment with 666 channel capability, the station class is 00. Refer to manufacturer information before changing this parameter. If a 666 channel phone is programmed with the wrong station class, the system may believe the phone can handle 832 channels. The result is excessive dropped calls and an upset customer.	**(Mobiles) 00 or 08**
SYSTEM ID: This five-digit decimal number identifies the customer's "home" system. The number is sent to the mobile by the cell for the mobile to figure out if it is in its home system. If the system ID sent by the cell doesn't match the phone's programmed system ID, the telephone's Roam light will display. Odd numbers denote system A and even numbers are used by the B carriers.	**XXXXX**
SYSTEM SELECT: See Selectable System Program	

	Standard
UNLOCK CODE: A three- or four-digit number used to unlock the phone. When the mobile is locked, outgoing calls cannot be placed. Depending on the model, incoming calls may not be allowed either. To unlock the phone, the user enters the three or four digit number from the keypad and the phone resumes normal operation.	**XXX**
VOX: A power-saving mode of operation for portables developed by MOTOROLA. For VOX operation, the phone must be equipped (and enabled when programmed) and the cellular system must be equipped. Those switches permitting VOX are generally limited to MOTOROLA. VOX is also known as discontinuous operation.	**0 or 1**

Appendix C Additional Resources

Cellular Telephone Directory

Communications Publishing
P.O. Box 500
Mercer Island, WA 98040-0500
(800) 366-6731 or (206) 232-3464

Everything you ever wanted to know about cellular roaming. The book is filled with system coverage maps, roamer access numbers, carrier telephone numbers and international roaming information.

Cellular Installation "THE BASICS" 101A — Cost $79.00

Bishop & Assoc, Inc.
P.O. Box 2027
Bothell, WA 98041
(206) 485-0572 or (206) 485-0572

A great training video for new installers and for persons in customer service, sales and other positions where an understanding of the installation process is vital. The video outlines and illustrates the installation process from vehicle check in, through component installation, testing and check out.

CELLMAPS

Cellular Directions, Inc.
P.O. Box 66843
Saint Petersburg Beach, FL 33736
(818) 345-6150

Cellular Directions offers roaming information in either format, books or disk. They also produce colorful wall maps that show system locations throughout the United States.

Digital Cellular Radio

Communications Publishing
P.O. Box 500
Mercer Island, WA 98040-0500
(800) 366-6731 or(206) 232-3464

This book has over 400 pages of great cellular information. It includes chapters on history, analog, digital and system design considerations. The book is a "7" on a technical scale of "10."

How To Win Customers

Bishop & Assoc., Inc. (or your local bookstore)
P.O Box 2027
Bothell, WA 98041
(800) 829-0572 or (206) 485-0572

The book outlines "win-win" customer strategies for building and retaining customers. It specifies five ways to keep customers coming back.

Inside Cellular

Communications Publishing
P.O. Box 500
Mercer Island, WA 98040-0500
(800) 366-6731 or (206) 232-3464

The author was vice president of sales and marketing at a successful cellular company. In the book, he describes the issues facing cellular management and offers solutions.

Installation Handbook — Cost $49.00

Bishop & Assoc, Inc.
P.O. Box 2027
Bothell, WA 98041
(800) 485-0572 or (206) 485-0572

Cellular installation tips, techniques and time savers. Loaded with information on specific vehicles, wattmeter usage, system operation and recommended tools and resources.

Moments of Truth

Bishop & Assoc., Inc. (or your local bookstore)
P.O. Box 2027
Bothell, WA 98041
(800) 829-0572 or (206) 485-0572

This book stresses the importance of each customer interface. Each "moment of truth" can make or break a business.

The Cellular Telephone Installation Handbook

Quantum Publishing, Inc.
P.O. Box 310
Mendocino, CA 95460
(800) 442-9666 or (707) 937-4488

The book is written by industry pro Michael Losee. It offers tips and techniques for every aspect of the mobile installation process. This book is also available from TESSCO and Communications Publishing. (ISBN 0-930633-05-9)

Periodicals

Cellular Business — Cost $24.00/Yr

Intertec Publishing Corp.
P.O Box 12960
Overland Park, KS 66212
(913) 888-4664

The industry standard of cellular magazines. It is published monthly and is loaded with all the latest cellular news.

Cellular Marketing — Free

Cardiff Publishing Company
6300 South Syracuse Way, Ste. 650
Englewood, CO 80111
(312) 762-2193

Cellular Marketing provides valuable marketing information for cellular agents and carriers.

Installation News — Free

Bobit Publishing Company
2512 Artesia Blvd
Redondo Beach, CA 90278
(213) 376-8788

This magazine is written for the cellular, autosound and alarm installer. It is filled with tips, ideas and suppliers.

RCR — Free

RCR Publications Inc.
1728 Downing Street
Denver, CO 80218
(303) 832-6000

RCR is published weekly and offers the breaking news about the cellular, two-way and paging industries.

Mobile Product News — Free

Phillips Publishing Inc.
7811 Montrose Road
Potomac, Maryland 20854
(301) 340-2100

Every issue has pictures and descriptions of the latest cellular, paging and two-way products. It's an easy way to stay product-wise.

Glossary Cellular Terms

ACTIVATION: The process of entering specific phone information into the network's data base. For a cellular phone to work, the system must be informed of the mobile's number and ESN. Once this information is entered, the phone can place and receive calls.

AMPS: Advanced Mobile Phone Service The term, originated by Bell Laboratories, describes the network and mobile standards.

ANALOG: A type of signal used with the first generation of cellular telephones. An analog signal can be monitored with a scanner and the conversation is understandable. The second generation of cellular phones uses a digital signal. Digital signals, if monitored, sound like the whir of a facsimile machine.

AUTONOMOUS REGISTRATION: A network feature designed to know the whereabouts of mobiles operating in the system. With autonomous registration, the switch intermittently requests idle mobiles to "sign in" with the system. If a mobile receives an incoming call, the system knows which cell the user is near.

BACK HAUL: A cellular network that shares a MTSO with a neighboring system is back hauling. The foreign switch is connected via microwave or telephone lines.

BOOSTER AMPLIFIER: This portable phone accessory will increase the phone's power output when operating in a vehicle.

CALL WAITING: A feature available with most networks to notify a user of a second call while an existing call is in progress. The user hears a tone, presses the SEND button and is connected to the second call. To return to the first call, the SEND button is pressed again.

CARRIER: The operating company offering cellular service is the local carrier. BELLSOUTH MOBILITY and CELLULAR ONE are examples of cellular carriers.

CELL: The cell is a component of the cellular system. Depending on a network's complexity, it may have many cells. The cell has transmitters, receivers and antennas. It is connected to the system's MTSO via microwave or telephone lines.

CGSA: Cellular Geographic Service Area. This is the actual coverage area for a cellular system. In other words, anywhere in the system where a customer's phone would go into service.

CONFERENCE CALLING: An option offered by most networks that allows users to connect with two called parties simultaneously. To start the feature, the first call is placed. Then the SEND button is pressed to set up the second call. The SEND key is pressed again and all three persons are involved in the same conversation.

CONTROL CHANNEL: The cells have two types of channels, control and voice. The control channels are used for call set up, autonomous registration and user notification. Control channels send and receive data only.

CONVERSION: This term is used to describe a customer who disconnects from one carrier and connects with the competing carrier in a cellular market. The cellular user is called a conversion customer.

CUSTOMER BASE: A network's customer base is anyone signed up with the carrier for cellular service. Fortunately, the entire customer base doesn't place calls simultaneously.

DATA: A type of signal transmission involving a non-analog method. Data is used on control channels and hidden on the voice channels. Data transmission sounds similar to a facsimile machine.

DATA CABLE: Data cables, generally found in mobile installations, carry the signal from the transceiver to the handset. In a vehicle, the data cable is routed under the carpet and behind panels to connect the phone's components.

DB: See Decibel

DECIBEL: A measurement used to describe a signals gain or loss. For example, an amplifier that takes a 1 watt signal and boosts it to 2 watts is called a 3db gain amplifier. If a signal is doubled, there is a 3db gain.

DROPPED CALLS: A cellular buzz word used to describe a call that was disconnected for no apparent reason. If the customer presses the END key, the call is ended. A dropped call is one that dies without warning to the customer.

DTMF: See dual tone multi frequency

DUAL TONE MULTI FREQUENCY: These are the tones that sound whenever the user presses a phone key. The tones are used to tell the system and phone accessories which number key was pressed.

ELECTRONIC SERIAL NUMBER: Each phone has a discrete factory programmed identification number. This electronic number is used by the system to verify that the correct phone is using a certain phone number. The electronic serial number (ESN) cannot be altered by anyone besides the phone manufacturer.

ELEVATED FEED: A type of antenna used for mobile installations. The elevated feed works well in locations lower than the vehicle's roof or areas with no flat metal surface.

ESN: See Electronic Serial Number

EUROPEAN SIREN: A tone occasionally heard by cellular users from the phone's handset. The European siren generally occurs if the phone's "No Service" light illuminates or the phone is defective.

EXPANDED SPECTRUM: The expanded spectrum was introduced to cellular in 1987. This increased the existing 666 available channels to 832.

FAST BUSY: A call denied by the cellular phone may be welcomed with a network recording or a fast busy. It is the system's way of telling the user that the call cannot be processed.

FCC: See Federal Communications Commission.

FEDERAL COMMUNICATIONS COMMISSION: This governmental agency oversees radio frequencies and allocations. It also is called the FCC.

FOCC: See Forward Control Channel

FORWARD CONTROL CHANNEL: Network control channels tell the user system information. The forward control channel is the signal traveling from the system to the mobile.

FRINGE AREA: Every cellular system has areas of "good", "bad" and "ugly" reception. The good areas are great. The ugly areas cause the phone's No Service light to stay on and the "bad" areas are very scratchy. The bad area is called a fringe area.

FORWARD VOICE CHANNEL: When a call is in progress, the mobile has jumped from the control channel to the voice channel. The mobile uses the voice channel during the conversation. The forward voice channel is the cell's channel sending the signal from the system to the mobile.

FVC: See Forward Voice Channel.

HOOK FLASH: A short tone originating from the mobile phone. The hook flash is also called signaling tone and occurs whenever the SEND or END keys are pressed.

HOOK SWITCH: See Hook Flash.

IMPROVED MOBILE TELEPHONE SERVICE: The bell system term for the mobile telephone service introduced in 1964. The acronym for this term is IMTS. IMTS was cellular's predecessor.

IMTS: See Improved Mobile Telephone Service.

INTERCONNECT: The network's method of connecting the cells to the MTSO.

KHZ: This abbreviation stands for Kilohertz. Kilohertz is used to describe a frequency or signal. Kilo means thousand and hertz denotes cycles per second. One kilohertz is 1,000 cycles per second.

LANDLINE: A "telephony" term describing phone line used by home or office phones.

MHZ: An abbreviation for the word megahertz. Mega means million and hertz stands for cycles per second. One megahertz equals one million cycles per second. This term is often used to describe a signal, i.e., cellular operates in the 800 Mhz range.

MIN: An acronym for the mobile identification number. This is the phone's ten-digit programmed number. The seven digit number plus the area code equals ten.

METROPOLITAN SERVICE AREA (MSA): Large areas with cellular coverage. An example of a MSA is Los Angeles or Miami.

MSA: See Metropolitan Service Area.

MOBILE TELEPHONE SWITCHING OFFICE: The brains of the cellular system. The cells and mobile phones are orchestrated by the commands of the MTSO. It also is called the MTX or Switch.

MTSO: See Mobile Telephone Switching Office.

NAM: The number assignment module, or NAM, is a component inside the telephone that stores the phone number and system parameters. NAMs are designed in reusable and disposable versions. A reprogrammable NAM is called an E^2PROM. This stands for electronically erasable programmable read only memory. Remember this definition and impress your friends.

NON-WIRELINE: The non-wireline carrier operates on the A channels.

RJ11: The type of jack or connection used with home phones. RJ11 jacks have 4 connectors and are used to connect the phone to the landline system. Your home phone plugs into the wall with an RJ11 jack. The RJ11 interface is used in cellular when connecting a landline accessory, i.e., facsimile machine, to a cellular phone.

ROAM: Whenever a cellular phone is operating on a non-home system, it is roaming.

RSA: This abbreviation stands for rural service area. An RSA is a smaller cellular system, usually less than 10 cells total.

RECEIVED SIGNAL STRENGTH INDICATOR: Usually a bar graph or meter to show how well the mobile receives the system's signal. A handy feature if the user suspects a poor service area and wishes to delay call placement until receiving a satisfactory signal.

RSSI: See Received Signal Strength Indicator.

REVERSE VOICE CHANNEL: When a call is in progress, the mobile has jumped from the control channel to the voice channel. The mobile uses the voice channel during the conversation. The reverse voice channel is used for the signal traveling from the mobile to the system.

RVC: See Reverse Voice Channel.

SAT: See Supervisory Audio Tone.

SAT COLOR CODE: The networks use three distinct SATs. A SAT's color code suggests the specific SAT; #1, #2 or #3.

SCC: See Sat Color Code

SID: See System Identification Number

SIGNALING TONE: A tone used by the phone to indicate requests to the system. This tone emits when the SEND or END keys are pressed. Also see Hook Flash.

ST: See Signaling Tone

SYSTEM IDENTIFICATION NUMBER: A discrete number used by each cellular system as a means of identifying itself to mobiles and portables operating within the network. The phone is programmed with a system ID. If the system's SID matches the phone's programmed SID, the roam light on the telephone will not illuminate.

SUPERVISORY AUDIO TONE: A tone used by the system to verify that the correct mobile is operating on a specific channel. It is used also by the system to figure out if a handoff is necessary.

VOICE CHANNEL: A channel in the cellular network used for telephone conversations. The other type of channel is called the data channel.

WATT: A unit of measure to denote power. A phone's power output is stated in terms of X watts.

WIRELINE: The B carrier in a specific market. The wireline operating companies are generally a derivative of the phone company.

YAGI: Yagi antennas are extremely directional. They are commonly found in stationary applications, pointed at the closest cell site.

Index

Index (Continued)

NOTES